AF775869

Science in performance

Manchester University Press

theatre
theory • practice
• performance •

series editors
MARIA M. DELGADO
MAGGIE B. GALE
PETER LICHTENFELS

advisory board

Michael Billington, Sandra Hebron, Mark Ravenhill, Janelle Reinelt, Peter Sellars, Joanne Tompkins

This series will offer a space for those people who practise theatre to have a dialogue with those who think and write about it.

The series has a flexible format that refocuses the analysis and documentation of performance. It provides, presents and represents material which is written by those who make or create performance history, and offers access to theatre documents, different methodologies and approaches to the art of making theatre.

The books in the series are aimed at students, scholars, practitioners and theatre-visiting readers. They encourage reassessments of periods, companies and figures in twentieth-century and twenty-first-century theatre history, and provoke and take up discussions of cultural strategies and legacies that recognise the heterogeneity of performance studies.

also available

Directing scenes and senses: The thinking of Regie
PETER M. BOENISCH

The Paris Jigsaw: Internationalism and the city's stages
DAVID BRADBY AND MARIA M. DELGADO (EDS)

Street theatre and the production of postindustrial space: Working memories
DAVID CALDER

Death in modern theatre: Stages of mortality
ADRIAN CURTIN

Theatre in crisis? Performance manifestos for a new century
MARIA M. DELGADO AND CARIDAD SVICH (EDS)

World stages, local audiences: Essays on performance, place, and politics
PETER DICKINSON

Foucault's theatres
TONY FISHER AND KÉLINA GOTMAN

Witness onstage: Documentary theatre in twenty-first century Russia
MOLLY FLYNN

Robert Lepage's original stage productions: Making theatre global
KAREN FRICKER

Performing presence: Between the live and the simulated
GABRIELLA GIANNACHI AND NICK KAYE

Queer exceptions: Solo performance in neoliberal times
STEPHEN GREER

Performance in a time of terror: Critical mimesis and the age of uncertainty
JENNY HUGHES

South African performance and the archive of memory
YVETTE HUTCHISON

Unlimited action: The performance of extremity in the 1970s
DOMINIC JOHNSON

Jean Genet and the politics of theatre: Spaces of revolution
CARL LAVERY

After '89: Polish theatre and the political
BRYCE LEASE

Not magic but work: An ethnographic account of a rehearsal process
GAY MCAULEY

'Love me or kill me': Sarah Kane and the theatre of extremes
GRAHAM SAUNDERS

Performing the testimonial: Rethinking verbatim dramaturgies
AMANDA STUART FISHER

Trans-global readings: Crossing theatrical boundaries
CARIDAD SVICH

Negotiating cultures: Eugenio Barba and the intercultural debate
IAN WATSON (ED.)

Science in performance

Theatre and the politics of engagement

SIMON PARRY

Manchester University Press

Copyright © Simon Parry 2020

The right of Simon Parry to be identified as the author of this work has been asserted by him in accordance with the Copyright, Designs and Patents Act 1988.

An electronic version of this book is also available under a Creative Commons (CC-BY-NC-ND) licence, which permits non-commercial use, distribution and reproduction provided the author(s) and Manchester University Press are fully cited and no modifications or adaptations are made. Details of the licence can be viewed at https://creativecommons.org/licenses/by-nc-nd/4.0/

Published by Manchester University Press
Altrincham Street, Manchester M1 7JA

www.manchesteruniversitypress.co.uk

British Library Cataloguing-in-Publication Data
A catalogue record for this book is available from the British Library

ISBN 978 0 7190 9920 5 hardback
ISBN 978 1 5261 5090 5 open access

First published 2020

The publisher has no responsibility for the persistence or accuracy of URLs for any external or third-party internet websites referred to in this book, and does not guarantee that any content on such websites is, or will remain, accurate or appropriate.

Typeset
by Sunrise Setting Ltd, Brixham

Science in performance

Theatre and the politics of engagement

SIMON PARRY

Manchester University Press

Copyright © Simon Parry 2020

The right of Simon Parry to be identified as the author of this work has been asserted by him in accordance with the Copyright, Designs and Patents Act 1988.

An electronic version of this book is also available under a Creative Commons (CC-BY-NC-ND) licence, which permits non-commercial use, distribution and reproduction provided the author(s) and Manchester University Press are fully cited and no modifications or adaptations are made. Details of the licence can be viewed at https://creativecommons.org/licenses/by-nc-nd/4.0/

Published by Manchester University Press
Altrincham Street, Manchester M1 7JA

www.manchesteruniversitypress.co.uk

British Library Cataloguing-in-Publication Data
A catalogue record for this book is available from the British Library

ISBN 978 0 7190 9920 5 hardback
ISBN 978 1 5261 5090 5 open access

First published 2020

The publisher has no responsibility for the persistence or accuracy of URLs for any external or third-party internet websites referred to in this book, and does not guarantee that any content on such websites is, or will remain, accurate or appropriate.

Typeset
by Sunrise Setting Ltd, Brixham

This book is dedicated to the memory of my sister Julie.

CONTENTS

LIST OF FIGURES

ACKNOWLEDGEMENTS

I did the research for this book over a long period of time and it has therefore benefitted from my relationships with a wide range of people, some of whom I mention here, but there are many more who have helped me practically, creatively and intellectually in a variety of ways, so thank you to you too.

I have a very supportive institution and colleagues at University of Manchester. The School of Arts, Languages and Cultures granted me a period of research leave in 2018 which enabled me to do much of the writing of the book as well as a small travel grant that supported research for Chapter 2. Kate Dorney and Jenny Hughes have been extremely generous in providing comments on drafts and being generally supportive and encouraging. Conversations with other colleagues have opened up new lines of thought at various points, through research seminars and a series of reading groups in theatre and performance organised by Steve Scott-Bottoms. Maggie Gale is always a source of wise advice.

I have been encouraged and stimulated by meetings of the Theatre and Performance Research Association Science and Performance Working Group and would particularly like to thank Alex Mermikides, Gianna Bouchard and Paul Johnson for their work setting this up and facilitating it.

Christina Runge was instrumental in giving me access to Gob Squad and *My Square Lady*, and also continued a helpful conversation with me about the project. Sean Patten, Manfred Hild, Sarah Thom and other members of the *My Square Lady* team were welcoming and thoughtful in discussing their work on the project.

Chapters 4 and 5 in particular emerge from my experience of working at the Wellcome Trust 2002–09. At the time, my thinking was challenged and opened up by many conversations with colleagues including Peter Finegold, Hannah Russell, Verity Slater, Meroë Candy, Marie-Lise Sheppard, Ian Jones, Rebecca Gould and Ralph Levinson. I subsequently collaborated on two phases of evaluation research on Theatre of Debate, commissioned by Y Touring, which directly inform the last section of Chapter 4, working with Peter Finegold (Isinglass Consultancy) and Jo Hutchinson (University of Derby), and then with Suzanne Parsons (Nowgen). Nigel Townsend has been a source of inspiration but also generous in commenting on drafts of Chapter 4. Many aspects of this book, but especially these chapters, benefitted greatly from the advice of my PhD supervisor, Helen Nicholson. My background understanding of public engagement with science and Sciart has been supported by conversations with Bella Starling, Ken Arnold, Siân Ede and Katherine Mathieson.

I'd like to thank Celia Moss and Matthew Walters for helping me rediscover *Colour Coded*; Charlotte Bennett and Nikki Barraclough for talking to me about *Genesis*; and Olivia Hirst for sharing material on *Goodstock*. Ayan Mahamoud, Said Jama Hussein and Martin Orwin gave me particular insights into *Khat Out of Sight Out of Mind*. Cynthia Hopkins kindly offered access to documentation of *This Clement World*.

Ceri Harrop and Lottie Clark provided helpful and enthusiastic research assistance at various points. Staff at Manchester University Press have been patient and efficient in producing the book.

Jenifer Parry and Colin Parry continue to inspire me through their sustained demonstrations of multiple ways of knowing. Big hugs to Rebecca Parry and Max Parry for your tolerance and love during the protracted process!

PREFACE

This book was written mostly during 2018, looking back across roughly the first two decades of the twenty-first century. I attempt to chart particular historical and geographical coordinates in the first chapter. However, my choice of focus on science in performance has a lot to do with my own desire to reflect on a range of practices I encountered in some form from 2001 onwards. At the turn of the century, I was involved in a range of work exploring the potential of theatre in intercultural communication and language learning. It was on the basis of this experience, reflecting widespread beliefs that cultural and disciplinary differences have some similarities, that I was employed in 2002 by the medical research charity, the Wellcome Trust, to develop a strand of work linking some of the charity's emerging support for science and arts collaborations with a particular strategy attempting to reimagine young people's science education for the twenty-first century. With my own interests in what was coming to be known as applied theatre, I was intrigued to explore what might happen when dialogical approaches to theatre and education met the hard facts of science.

In many ways this was a slightly daunting proposition. I had no particular science educational background. I had some sense of the geographies and histories of young people's theatre in the UK, but not all of my overtures with different companies and practitioners working in this field went well. Some of those with backgrounds in theatre-in-education, theatre of the oppressed or other related practices were suspicious of offers of collaboration or potential funding for a combination of understandable overlapping philosophical and ideological reasons. This had partly to do with concerns about instrumental use of the arts in general within education. There were associated fears that accepting funding from a medical or scientific charity might come with pressure to make work that was 'didactic' or 'message-driven'. Some had suspicions about the corporate agenda of the Wellcome Trust set up from the will of Sir Henry Wellcome and linked in many people's minds with the pharmaceutical companies that also carried his name.

I was reassured though by the precedent that already existed. One of the first companies I met, Y Touring, discussed at length in Chapter 4, had both strong roots in theatre-in-education and also an already impressive track record opening up complex scientific ideas through theatre. The Wellcome Trust had been funding the arts since 1996, and although the Sci-art funding schemes had reached a rather different community of practice, dominated by the visual arts and not necessarily as concerned with dialogical processes as applied theatre or theatre-in-education, some of these projects had also turned to theatre or performance.[1] In 2002, Wellcome also supported a young people's UK national science and theatre project *Science Centrestage* in which school students worked with drama practitioners and science advisors to devise original performance pieces. Many of these performances established new connections between science and drama teachers within schools and stimulated diverse responses to a range of scientific themes. The work produced by the young people as part of this festival was in no way adversely affected by its scientific content. If anything, some of the work seemed to have risen to the challenge of complexity in producing work that was original and moving.

Between 2002 and 2009 when I worked at the Wellcome Trust I repeatedly encountered work that emerged in surprising, moving and sometimes disturbing forms. There are many examples that offered new perspectives and I have not the space here or in the rest of the book to include them all.[2] Certain images lodged in my head though and have propelled work on this book whether or not they appear in the final version. In *Visiting Time*, Anna Ledgard and Mark Storor haunted a Dorset hospital with a group of pyjama-clad, pillow-headed children in response to a young man's story of living with cystic fibrosis.[3] *Stan's Cafe* made school students into performance epidemiologists mapping population health by filling school halls with carefully measured piles of rice.[4] A group of MCs, whom I will return to in Chapter 5, rapped on a Birmingham rooftop mixing dermatology with grime. A drama school organised a group of young trans people in a process that was part artwork, part research project, part activist intervention.[5] The scientists, artists and educators working on these projects and others often had to forge practices and processes that, to some extent, defied the political and aesthetic frameworks I had to work with at the time. This book is partly an attempt to formulate such frameworks retrospectively.

The book is not, however, about defining a set of practices, but rather explores political and aesthetic questions that are prompted by them. I am interested in the potential such practices might hold for new ways of knowing and being in the world that might be derived from science in performance. While the projects just outlined and the work discussed in Chapters 4 and 5 of this book explicitly engage with the discourses and policies of public engagement with science and science education, the potential of theatricality to connect knowledges stretches beyond these contexts. In Chapters 2 and 3, I therefore extend my methodology – perhaps beyond breaking point – to explore how performance practices articulate (scientific) knowledge across a range of intersecting traditions including opera, comedy, contemporary theatre and the musical. Chapter 2 tracks a collaboration between robotics researchers, a major German opera house and the performance collective Gob Squad. Chapter 3 looks across a range of examples of popular theatre exploring the ways they reflect and shape engagements with contemporary science. In the next chapter, I will outline the broader political context for engagement between science and performance and explain why theatricality may offer ways to make sense of this engagement.

Notes

1 For accounts of work funded through the Sciart programme see Arends and Thackara (2003); for further details of theatre, performance and film funded by Wellcome in the early years of its arts funding see Arends and Slater (2004).

2 There are two publications documenting a range of these practices (Levinson *et al.* 2008; Parry 2005).

3 Since *Visiting Time*, Ledgard and Storor have developed a series of groundbreaking, imaginative projects that bring together medical perspectives with the stories of patients and carers in surprising and deeply moving forms. For more detail see Ledgard's chapter in *Creative Encounters* and other sources (Levinson *et al.* 2008; Nicholson 2011; Walsh and Ledgard 2013).

4 *Plague Nation* was a bespoke school-based version of the company's hugely successful touring piece *Of All the People in All the World*. For more details see the company's website and other sources (Crossley and Yarker 2017; Parry 2010).

5 The project run by Jay Stewart and Catherine McNamara was an early experiment by the now highly successful organisation Gendered Intelligence: http://genderedintelligence.co.uk/.

1

Science in performance: convergence, emergence and divergence

Starting with a (big) bang

> Sir Ian McKellen as Prospero: Miranda, go out into the world. Will you be for all of us gathering here our eyes, our ears and our hearts? Shine your light on the beautiful diversity of humanity. Understand those rights that protect us. Look up, stretch your wings and fly. Will you take the journey for all of us and will you set us free?
>
> Professor Stephen Hawking: We live in a universe governed by rational laws that we can discover and understand. Look up at the stars and not down at your feet. Try to make sense of what you see and wonder about what makes the universe exist. Be curious. (Paralympic Games 2012)

A golden, smoking orrery inscribed with 'human rights for all' descends into a giant umbrella that erupts with pyrotechnics. Dancers hold umbrellas that change colour under the lights. They spread out from the centre and move into a dance sequence to Rihanna's hit song *Umbrella*. Aerial performers spin over their heads and street dancers from Flawless perform acrobatic moves on stage beneath a giant model moon. The Paralympic Games Opening Ceremony at London 2012 started with a/the Big Bang. Prominent figures of theatre and science in the UK, McKellen and Hawking, uttered the lines quoted above in their distinctive tones.

McKellen spoke in character as Prospero to Miranda, played by Nicola Miles-Wildin, within the loose narrative of *The Tempest* that was threaded through the spectacle.

The spectacle, entitled *Enlightenment* and directed by Jenny Sealey and Bradley Hemmings, followed Danny Boyle's widely acclaimed Olympic Opening Ceremony and garnered comparable praise in the media. Its title links science, culture and politics within a particular historical narrative. Science is conceived as a set of rational laws but also as a process of discovery, observation and understanding driven by passions such as curiosity and wonder. Scientific enlightenment is juxtaposed and inextricably connected to an appreciation and understanding of diversity and human rights discourse. Miranda, representing all of humanity, searches for freedom with two of her senses, according to Hawking's injunction, but also with her heart. Miranda's sense beyond sensory perception though is foreshadowed in Hawking's introduction when he asks, 'what is it that breathes fire into the equations, and makes a universe for them to describe?' What is this fire? Why does the universal potential of maths need to be ignited with energy from beyond the senses? If the magical world of *The Tempest* provides the universe of the spectacle, where Miranda can grow wings and fly off in her search for freedom, what is Hawking doing there telling us/her about another universe?

This moment, broadcast worldwide to an audience of many millions, is a suitable place to start this book as it introduces its main themes and raises a number of its key questions.[1] The book considers how and why science provides inspiration for and interference in contemporary performance across a range of contexts from mass global spectacle, to theatre stages, school classrooms and hospital wards. It asks questions about the way these performances represent science as laws, process, method, vocation or otherwise. It explores how and why science interrupts or intervenes in theatre and vice versa. It also reflects on the politics of the breaking apart or coming together that these events might constitute. That is to say, it considers the way that science in performance assembles people and things in particular ways.

For all my interpretive suspicions, based on critiques of spectacle in general and the particular globalising and commodifying tendencies of events such as the Paralympics, I found this moment in the performance and indeed the dance sequence described above strangely moving. Indeed, the spectacle is in constant motion: on wheels, through the air, dancing, running, crawling, leaping across the ground. What was driving this (e)motion and where was it heading?

Hawking's account of Big Bang theory and the associated idea of 'eternal inflation' suggest the cosmos is expanding (Hawking and Hertog 2018).

Galaxies or even universes diverge with increasing dark energy between them. Hawking's briefly stated philosophy of science offers diverging possibilities for us to make sense of this condition. Look down at my feet *or* up at the stars? Make sense of what I see *or* wonder about what makes the universe exist? Hawking's monologue does seem to make sense but only because his rhetoric and the aesthetic of the performance more broadly can hold together these apparently divergent demands on perception. The staging and the choreography set up a series of juxtapositions: Hawking is next to McKellen, the boundlessness of human endeavour is counter-posed to the boundlessness of the universe. Prospero/McKellen's humanist injunctions diverge too. Can Miranda be all of our senses *and* shine a light on all diversity? Will she take the journey for us *or* set us free? Do these conjunctions set up oppositions that can be reconciled or stark alternatives? The scene seems to aspire to convergence both within and between Hawking's vision of science and Prospero/McKellen's humanism. The staging connects them in body and speech. It attempts to reconcile the divergences, even as it acknowledges them, with the convergent desires of the Paralympic movement, science's universal laws and universal human rights.[2]

Miranda/Miles-Wildin, often suspended right above the action, is both the central character but also the central spectator for the artistic representations of science that form large parts of the performance including music inspired by Isaac Newton, technical defiance of gravity, animated human images of the Big Bang, the Solar System and the Large Hadron Collider. If Prospero is doubled in the Paralympics Opening Ceremony in the figures of Hawking and McKellen, Miles-Wildin seems to stand in for Ariel, Ferdinand and Miranda as agent of and audience for Prospero's visions. Much of the structuring dialogue of the piece is spoken by Prospero to Miranda. He directs her and our attention to each of the scenes as they unfold. She models a curious, attentive, engaged public for lessons about science, rationality, beauty and human rights that constitute the enlightenment of the title. She manifests this engagement as she speaks the famous lines from *The Tempest* that, in this performance, seem to allude to the way the Paralympics makes visible human diversity:

> Oh, wonder!
> How many goodly creatures are there here!
> How beauteous mankind is! O brave new world,
> That has such people in't!

These lines also, of course, hold fearful resonance as the source of Huxley's dystopian 1931 vision of a *Brave New World* (2007) in which technology resources social control, hierarchy and suppression of difference. Such mixed up feelings of science in performance are a challenging undertow to the generally celebratory atmosphere of the spectacle generated by the often implicit allusions to the experience of marginality and discrimination of disabled people historically and today.

The playful imagery of the opening ceremony juxtaposes the universalist iconography of global institutions with symbols associated with British national identity. The Paralympics' own iconography frames the piece and the Universal Declaration of Human Rights is written across Miranda's shirt and wheeled on as a giant red leather-bound tome. The Big Bang, umbrella and orrery very loosely constitute objects emerging from British science and technology in the twentieth, nineteenth and eighteenth centuries respectively.[3] These objects represent a human history of scientific and technological progress from a parochial subject position. Opening ceremonies of major sporting events are globalising acts of emergence with invariably normative, nationalist aims. They are read by both domestic and global audiences as constitutive of a particular moment in a national identity even as they orient themselves towards an international public. In this staging, the umbrella is a metaphor for the internationalist aspirations of human rights and the Paralympic movement, a technology for sharing shelter from the stereotypically British weather, a symbol of colonial empire and, in the global hit song by a Barbadian popstar, a technology of postcolonial empire. Yet these objects also appear to gesture beyond the human: to an emergent history of the universe where humanity makes a brief appearance; to climatic forces with which we are painfully at odds; to a cosmology in which humans are minutely peripheral. The tendency for theatrical performance to emerge from the political paradoxes of science runs throughout the examples discussed in this book.

Theatre, as incarnated in the figures of McKellen/Prospero and Miles-Wildin/Miranda, struggles with the decentring of the human subject even as it attempts to grapple with the emergence of new objects; science through the figure of Hawking seems to insist on it. Science introduces objects to the human world and, through its development of new technologies, has also transformed the planet. Science presents the human with the nonhuman, often by means of performance, and at the same time seems to be contributing to the destruction of the human by the nonhuman: hastening a post-Anthropocene age. This book examines the way that theatre and performance contribute to, reflect on and intervene in these convergent, divergent and emergent processes. Theatrical performance holds potential not just for supporting wider popular understanding of or engagement with scientific knowledge. Theatre might also contribute to opening up, for scrutiny, intervention or challenge, the processes and practices that go to make this knowledge, give it authority and thereby shape the future.

At one point in the ceremony, Stephen Hawking's voice is mixed into Orbital's electronica track *Where is it going?*

> The Large Hadron Collider at CERN is the largest, most complex machine in the world, possibly the universe. By smashing particles together at enormous energies, it recreates the conditions of the Big Bang. The recent discovery of what looks like the Higgs particle is a triumph of human endeavour and international collaboration. It will change our perception of the world, and has the potential to offer insights into a complete theory of everything.

As Hawking speaks, performers holding red umbrella-like particles transform the stadium into the Large Hadron Collider. Abruptly, the rhythmic chords and beats break into Graeae Theatre's version of *Spasticus Autisticus* by Ian Dury and the Blockheads, originally banned by the BBC, sung here by John Kelly from the cast of the stage musical *Reasons to be Cheerful*:

> I'm spasticus, I'm spasticus
> I'm spasticus autisticus.

It is an uneasy transition echoing the statements at the start from McKellen and Hawking but without their apparent rhetorical cohesion. For a moment it is jarring and in your face. According to Alfred Nordmann, the science of the Large Hadron Collider belongs to the 'age of science' in which huge resources go into the maintenance of the separation between intervention and representation in the face of the pressures of technoscience.

> As opposed to the 'sciences' (as conceived, especially, by scientists and philosophers of the nineteenth and twentieth centuries), the 'technosciences' do not even attempt to distinguish between theoretical representation of the world and technical intervention into the world. (Nordmann 2010, 7)

The Large Hadron Collider is a major collective human effort to hold intervention and representation apart in the production of facts. Hawking's commentary and indeed the concept of the spectacle as a whole invoke what Nordmann (2010, 12) calls the 'mythical alliance of science and the Enlightenment' in which science working towards a complete theory of everything is orientated towards a future conceived as 'a state of truth, justice and beauty'. *Spasticus Autisticus* is a war cry that, for one theatrical moment, rather than expose enlightenment's futures as myth, sets up its own mythology. Dury wrote the song as a response to the UN declaring 1981 the International Year of Disabled Persons. It was an act of cultural resistance to either representation or intervention and a statement of solidarity. 'I'm Spasticus' references the well-known line from the Kirk Douglas film *Spartacus* (1960) about the slave rebellion against the Roman Empire. The song was also an act of reclamation of the term 'spastic' that originates in scientific medical discourse but was often used insultingly to refer to disabled people such as those living with neurological conditions like Cerebral Palsy. For a moment here the theatrical spectacle tenuously holds together an emergent form of mythical alliance between Spasticus Autisticus, the Higgs particle and the Large Hadron Collider. It is an alliance forged from the human and the nonhuman, between a mythical figure of resistance derived from the use and abuse of biomedical knowledge, a particle of matter with divine aspirations, and a human institution built to provide universal and universalising access to otherwise imperceptible matter. Perhaps idiotically, this book places some faith in such an alliance.

Science in performance is not hard to find however it is conceived. The 2012 Paralympic Games Opening Ceremony is just one of many possible examples that seem to reflect a turn to performance by science and a reciprocal turn to science by theatre-makers. This chapter situates the practices discussed in this book in relation to their wider historical and geographical contexts, as well as in relation to more precisely located accounts of public engagement with science and other discourses of interdisciplinarity. It also reflects on the methodology I have adopted in response to this context and the questions prompted by the practices I have encountered.

Science, society and the pressure to perform

The institutions and practitioners of science feel the pressure to perform as much as, if not more than, any other field of activity. Steven Hawking's participation in the 2012 Paralympic Opening Ceremony exemplifies one way that many individual scientists have responded to this pressure through more or less engaging modes of public performance. Universities, research institutes and private companies increasingly value and audit this performance work of scientists as an ancillary or even predominant dimension of their practice. Since the 1970s, in the UK and the USA, scientists such as Richard Dawkins, Carl Sagan and Susan Greenfield have forged careers in which their skills and profiles as public performers became inextricable from their scientific training and identity (Fahy 2015). These celebrity scientists have an ambiguous relationship to the public based on the authority of their credible, scientific knowledge but also on the more nebulous allure demanded and generated by mass media performance. They seem to blur the boundaries between populism – an engagement with, against or otherwise in relation to popular structures of feeling – and what the sociologist Robert Merton (1973) had identified in 1942 as a distinct normative ethos and institutional structure of science.

At the same time, the pressure to perform is not just manifested in individual scientists' work as public performers. As the performance theorist Jon McKenzie (2001) has argued, this pressure spreads across a range of fields of activity. Performance management pervades the way organisations manage their employees. Novel devices and scientific theories perform (or fail to perform) as technology. McKenzie goes so far as to argue that performance actually works to displace discipline through a paradigm shift but one in which the paradigm is strangely emptied of meaning. To perform just means to perform. Disciplinary categories of truth are thereby disrupted leading to what some commentators in the media have called a post-truth era (d'Ancona 2017; Davis 2017). A 'crisis of truth', apparently brought on by the circulations and manipulations of emotion and belief within mass and now social media, seems to lead to perceptions of a 'crisis of trust' in the possibility of objective positions (Boler and Davis 2018). This initially threatens belief in the objectivity of news media but extends to other realms in which special claims to truth and objectivity are made. Science and scientists, even if not the first or even primary targets, are particularly sensitive to such threats. It is no wonder therefore that the performance principle, whether manifested in its organisational, cultural or technological modes, is met with ambivalence by many attached to the sciences and their institutions.

Despite varying levels of enthusiasm from scientists and scientific institutions, pressure on them to converge with other spheres of activity under the performance principle has been applied from a number of directions. There is a long history of this convergence that can be understood in terms of performance pressure. In a landmark text, historians Steven Shapin and Simon Schaffer (2011, 60) argued that particular social technologies were integral to the origins of modern science. These included what they call 'virtual witnessing' whereby the knowledge produced through science is validated by a larger public than has directly observed any of the experiments. Such a 'technology of trust and assurance' relied on the emergence of institutions such as the Royal Society founded in 1660. Many of the national institutions of science and discourse, through which relationships between science and society are conceived and practised in the twenty-first century, emerged during the Second World War and from under the shadow of the atomic bomb. During the war, research and technology development in the major combatant nations (USA, UK, Germany, Japan and the Soviet Union) was aligned and organised to serve the military interests of the respective nation-states. Science and society, even under the different political regimes of the Cold War, have been, in the terms of scholars of science and technology, 'co-produced' (Jasanoff 2013). This state of affairs – or as Sheila Jasanoff puts it 'states of knowledge' – largely persists, albeit with qualifications and modifications resulting from limitations to state power under globalisation. A canonical theatre history of science policy in the second half of the twentieth century charts this well, bookended by Brecht's post-war rewrites of *The Life of Galileo* (2001) in response to the devastation of Hiroshima and Nagasaki, and then by Michael Frayn's *Copenhagen* (1998). Both these plays turn and return dramatically to revolutionary moments in scientific history sketching through their dramaturgies complex reciprocal relationships between conceptions of the physical world and modes of socio-political organisation.

This iterative process of co-production is continually rehearsed and reframed: as the 'military-industrial complex' by President Dwight Eisenhower in his farewell speech in 1961 and then in Peter Drucker's notion that, since the Second World War, the USA had transformed from 'an economy of goods' into a 'knowledge economy' (1971, 263). This notion subsequently became a matter of consensus among major political leaders around the world even as it fuelled international competition. In Bill Clinton's Fifth State of the Union Address (1997), he outlined an education policy to equip US students 'to succeed in the knowledge economy of the twenty-first century' citing the importance of access to the internet. The Chinese president Jiang Zemin stated, a year later, that member states of the Asia-Pacific Economic Cooperation needed to develop new technologies to keep pace with the unfolding knowledge economy (1998). British Prime Minister, Tony Blair, argued in 2000 that he aspired for 'Britain to be at the forefront of the knowledge economy' going on to target universal internet access by 2005 (The White House Office of the Press Secretary 2000). Over a decade later, Indian Prime Minister Manmohan Singh was still setting a 'focus on investing in the knowledge economy' (2012).

Promotion of the knowledge economy, alongside the apparently uncontested conception of the knowledge that resources it, seems almost ubiquitous in political rhetoric. However, this sense is of course superficial. At the same time as the knowledge economy is rhetorically performed into being, diverse critics have argued for greater convergence or clearer separation between scientific knowledge production and its applications. It is notable that two figures who had been at the heart of collaboration between science and the military during the Second World War occupied key roles in immediate post-war US science policy. The former general turned President Eisenhower and the head of the wartime Office for Scientific Research and Development Vannevar Bush both articulated concerns after the war about too close a relationship between the institutions of science, private industry and military state power. Bush attempted in the immediate post-war period to insulate certain kinds of science through the notion of 'basic research' and via the institutional protection of a National Research Foundation. The extent to which he succeeded in institutionalising a new relationship between science and the state is contested. Science policy historian Michael Dennis has argued that the wartime organisation largely perpetuated in spite of Bush's efforts. Eisenhower's 1961 speech and the continued dominance of the US military in terms of federal spending on research and development seem to support this claim (Dennis 2013; Kennedy 2012).

More recent rhetoric in science policy tends to downplay military applications in favour of promoting performance in terms of the economic potential of convergence between the sciences and social policy. They also tend to be framed as driven by emerging science and technology rather than public policy aims. Roco and Bainbridge (2002, ix) proposed a particularly ambitious project to 'enhance human performance' in a report to the US National Science Foundation and the White House, which both describes and promotes convergence between nano-, bio-, information and cognitive sciences 'based on material unity at the nanoscale and on technology integration from that scale'. The convergence advocated here is between scientific fields, between science and technology, and between science and society in the ultimate, unified technoscientific performance. It realises technoscience – so the co-production of knowledge and technologies – as an apparently coherent and coordinated programme.[4] Such technoscientific programmes in their various forms have been subject to systematic critique from liberal, feminist, postcolonial, global justice and queer theoretical perspectives. These different perspectives have exerted their own performance pressure, emerging from and giving rise to a variety of performative practices and political actions.

The liberal democratic mode of science in performance is well exemplified in the UK through a series of attempts to connect science and publics through policy discourse and associated reforms in research funding. Despite the emergence of celebrity scientists in the mass media and an increasing variety of popular science formats, science policy has manifested a persistent concern with creating or influencing relationships between science and publics. These policy developments have contributed to pressures on scientists and scientific institutions to perform publicly and have supported the emergence of sets of practices known under various terms, including: public engagement with science, science communication, science outreach, citizen science and public involvement with research. The Royal Society's (1985) report, known as the Bodmer Report, articulated discourse and promoted practices associated with public understanding of science. It promoted increased and improved science communication by scientists as well as advocating changes in science education and science coverage in the media. The report argued that increased understanding of scientific facts and methods would lead to both economic prosperity and better private and public decision-making. Fifteen years later, the relationship between science and society was seen as suffering from a 'crisis of trust' and public 'ambivalence' rather than (or in addition to) lack of understanding, as reflected in a report on *Science and Society* to the House of Lords Select Committee on Science and Technology (2000). This report shifted the terms of the practice to that of the more emotionally invested public engagement with science. It registered the complexity of public attitudes towards science and reflected on the proposition that increased understanding might not lead to increased acceptance or appreciation of science. Such a shift in political thinking inflected science's turn to performance and influenced its theatrical strategies.[5]

The 'crisis of trust' articulated by the *Science and Society* report was linked to specific events or issues from the last decades of the twentieth century and to some extent, specifically located in the UK or Europe. The Bodmer report had been swiftly followed by the Chernobyl disaster and its fallout across Europe, and the 2000 report seems to respond directly to anxiety around food safety, by that stage also fuelled by the BSE/CJD crisis affecting the UK throughout the 1990s. The pressure generated by these controversies led to iconic moments of performance, as in 1990 when the then Minister of Agriculture, John Gummer, attempted to feed his 4-year-old daughter a beef burger to convince consumers of the meat's safety. In a subversive move prefiguring future eco-activist practices, she refused to take a bite. The report also reflects on UK and European public concerns over developments in the biosciences, notably the emergence of technologies linked to genetics including modification and cloning of living organisms.

The report is situated in time and place by its relative focus on biosciences and away from the risks of nuclear war (if not nuclear accidents) that dominated public cultures and performances of science during the Cold War. However the discourse in the report and the perceptions of public concerns it responds to have been more widely reflected in slightly different ways around the world. The genetically modified Flavr Savr tomato was approved for human consumption by the US Food and Drugs Administration in 1994 and the cloned sheep Dolly was born at the Roslin Institute in Scotland in 1996. Genetically modified organisms became an object of more global concern as Bt (bollworm resistant) cotton was approved first in the USA, then China and then, most contentiously, by the US corporation Monsanto in India in 2002. Many of the practices discussed in this book have been accounted for in the science policy terms of public understanding or engagement by employers (of scientists who participated), funders, or indeed audiences. These practices all articulate with the performances of understanding or engagement in complex ways that I will discuss in detail in subsequent chapters.

Public understanding and engagement as an evolving movement positions science as co-produced with or by institutions of civil society. Indeed, specific new civil society organisations grew out of and along with the policy discourse and helped to shape it. The London-based Science Media Centre was set up in 2002 as a direct consequence of the 2000 report operating in relation to a philosophy that: 'The media will do science better when scientists do the media better' (Science Media Centre n.d.). The scientist performers mentioned at the start of this chapter are exemplary practitioners of the kind that the Science Media Centre aspires to proliferate. The Science Media Centre's move from the Royal Institution to Wellcome Collection also seems to reflect the way public engagement marks a renewal and reform of the institutions of public, democratic or liberal science.[6] There is a comparable influence on US science policy through the work of independent charities such as the Howard Hughes Medical Institute and the Bill and Melinda Gates Foundation. All these charitable trusts, of course, were endowed from profits derived from the commercial development and mass-marketing of novel technologies.[7] The trusts have themselves now become significant actors in science policy.

Looking back from 20 years later, there are also notable and perhaps revealing absences from the 2000 report though. There is little significant reference to climate sciences or technologies of artificial intelligence and cognitive enhancement. These emerged as issues of widespread public concern in the 1990s. The United Nations Framework Convention on Climate Change was signed in 1992 and the Kyoto Protocol in 1997. The IBM computer Deep Blue famously beat the chess grandmaster Gary Kasparov in a six-game match in 1996. Despite the proportions of the global pandemic, in a decade during which new infections peaked at approaching 3.5 million and global deaths were still rising annually, there is also a strange absence of reference to HIV/AIDS.

While these issues were no doubt of widespread concern to large sections of the public, they perhaps sat at an oblique angle to the performance pressure of the public understanding or engagement agenda driven by and driving the House of Lords report. The convergent practices responding to awareness of climate change and its human causes, as well as cultural responses to the HIV/AIDS pandemic, were at this stage perhaps regarded as primarily social problems or rather too closely associated with activist movements for some of the liberal institutions of science (communication) to readily align themselves with. This book examines a range of practices that I argue sit on a blurred boundary between institutional (convergent) public engagement and activist (emergent) performance practices. These practices co-produce knowledge often in collaboration with scientific and cultural institutions but also develop their own social structures through creative processes. The particular areas of scientific knowledge addressed, notably biomedicine, climate sciences and artificial intelligence, reflect the major socio-scientific concerns of the era. Through the performance practices discussed though, these national and international concerns are relocated in relation to the particular stories and locations of diverse groups.

Cosmopolitics, theatre and idiocy

The somewhat partial focus of national science policies, as exemplified in the 2000 report, and indeed the sometimes narrow outlook of the institutions of civil society, has prompted a series of systemic political critiques of technoscience. These critiques can be located within critical science and technology studies, ecofeminist theory, political ecology, theories of global cognitive and epistemological justice, and queer theory. Such theoretical accounts are interwoven with politics of co-production exemplified in histories of eco-activism and indigenous peoples, women's groups, alternative globalisation movements and patient-driven health activism. The politics and forms of these co-productions can be read as and through performance and provide an alternative context for this book to the history of liberal institutions of science (communication) outlined above. They mark shifts from the distinct subject positions of the Mertonian scientist, the dis- or un-interested and therefore potentially disengaged citizens of liberal democracy or the technoscientific consumers of convergences between nanotechnology, biotechnology, information technologies and cognitive science.

For example, in his book *French DNA*, the anthropologist Paul Rabinow (2002) has provided an influential account of the way human genetics in France was co-produced with and by a patient organisation, L'Association Française pour la Myopathie (AFM), advocating for genetic research into rare inherited conditions, and a scientific research institute, Fondation Jean Dausset-CEPH (CEPH). Performance pressure here is exerted through novel manifestations of all three of McKenzie's (2001) modes of performance: the performance management exercised by the patient organisation in relation to the research institute that organised networks of scientists and patient activists in a form that Rabinow calls 'biosociality'; the particular technological performance resulting in the sequencing of specific rare disease genes by the CEPH and treatment-focussed research and development; and the cultural performance of the annual 'téléthon' fundraising event run by the AFM that underwrote the research efforts. Rabinow identifies the significance of this case in the emergence of a particular form of political subjectivity through the patient groups. He also argues that the 'mystical benevolence' of these groups presents an implicit challenge to technoscience and shifts the dominant conception of life itself. Mysticism, myth-making, forms of spirituality or even magic emerge in performance at various points of this book. Such phenomena are symptomatic populist counter-reactions to and modes of engagement with modern science in performance that reflect the complex feelings of particular groups. All of the artistic practices discussed in the second half of this book emerge from more or less novel forms of biosociality including alliances between individuals and groups identifying variously as artists, activists, scientists, patients, carers and indeed other positions. The challenge many of these practices attempt to address are both the feelings within patient groups that might be articulated in apparently non-scientific terms, of mysticism or spirituality for instance, but also the defensive, aggressive or indignant response this might provoke within scientific institutions or individual scientists.[8]

A related and influential history of co-production between health activists and scientific research can be located in relation to HIV/AIDS. A more antagonistic relationship between different political actors and the global spread of the pandemic has resulted in a much wider range of organisational, technological and cultural performances than in the case of *French DNA*. Organisations such as ACT UP, that started in the USA but then expanded into an international network, and Soul City in South Africa have, like the AFM, evolved through modes of organisational, technological and cultural performance (Epstein 1996; Gould 2009; Roman 1998). In this case, emergent biosociality overlapped with and to some extent was determined by pre-existing power dynamics. ACT UP mobilised politically through its close identification with the gay rights movement. The movement's challenge to heteronormativity underpinned and resourced its critique of technoscience. Soul City has had to negotiate the particular politics of HIV/AIDS in South Africa interwoven as it is with global health inequalities and the legacy of nationally institutionalised racism.

Challenges to technoscience from the perspective of the global south are controversially theorised within eco-feminist thought. In an iconic publication of the 1990s, Shiva *et al.* (2014) attempt to connect a feminist critique of patriarchy with a critique of globalisation from the perspective of the global south. In this work they link the operations of a 'capitalist patriarchy' with an 'ideology' of modern science. For them, the disregard for women's and indigenous knowledges are co-extensive with the disregard in materialist models of production for women's historical labour in reproduction, regeneration and subsistence (as outside production). For them, modern science is complicit with and generative of a knowledge hierarchy characteristically placing man before woman, white over black and culture over nature. Scientific knowledge production is based on 'violent separation' and tends to reductionism through 'partiality' – the exploitation and dissection of nature and the taking of parts for wholes (Shiva *et al.* 2014, 47). Shiva, who holds a PhD in Physics, has advocated for an alternative ecological scientific practice that resists knowledge hierarchies recognising and not exploiting indigenous and women's knowledges.

My perspective in this book attempts to draw on elements of such critique, and indeed on scientific ideas, without adopting either wholesale as a political or philosophical perspective. This decision is partly practical. The level of detail required to understand the politics of knowledge production and evaluation within a particular case, as conducted in many science studies texts, would have pulled the focus away from examining a diversity of theatre and performance practices. It has also not been within the capacity of this book's research to explore the way knowledge asymmetries have been challenged across and around the world. However, I do discuss relationships with these asymmetries as they are manifested in different socio-cultural contexts. More importantly though, while these critical perspectives are important, this book is less interested in the way theatre or performance might conduct a critique of science and more interested in the way theatre, performance and science might trace new associations between humans and between humans and nonhumans. The historical context sketched out above suggests a variety of manifestations of perceptions that knowledge is broken or that there are serious ruptures between knowledge-making practices.

Accounts of these ruptures are numerous. An impression of knowledge as broken emerges from the science wars discussed here below and from critiques of technoscience from a global justice or feminist perspective (Haraway 1997; Merchant 2006; Santos 2014; Shiva *et al.* 2014).

My intention throughout this book is therefore more one of repair than critique. Both my analytical perspective and indeed the choices of practices to explore are driven by an attention to the reparative potential of artistic practices. This approach follows a parallel trajectory located in the work of scholars in science studies, notably Isabelle Stengers and Bruno Latour, and also in the work of the US queer theorist Eve Kosofsky Sedgwick. Sedgwick argued for a practice of 'reparative reading' to replace or at least counter-balance what she called 'paranoid reading'. As important as demystification might be both in terms of the social and physical substrates of matter, being in such a demystified state does not inherently enable ethical action or the production of a just social order. As Sedgwick admits, it may seem like a common-place assertion to say 'that knowledge *does* rather than simply *is*'. However the unintended consequence of the critical and scientific preoccupation with demystification 'may have made it less rather than more possible to unpack the local, contingent relations between any given piece of knowledge and its narrative/epistemological entailments for the seeker, knower, or teller' (Sedgwick 2003, 124). Sedgwick is not arguing for naivety or a denial of reality, but rather for a broader range of positions or 'relational stances' within practices of knowing. A core argument of this book is that theatre can offer an experimental realm for exploring different stances in relation to emergent objects. Such stances might enhance or repair existing scientific or social accounts.

Paranoid reading in its mobilisations of 'strong theory' and 'negative affects' connects the repeated rehearsals of what has often, rather unfortunately, been called the science wars. Such 'wars' have been played out at various levels of bellicosity by Arnold and Huxley in the nineteenth century, Snow and Leavis in the mid-twentieth century, and Sokal and his various (mostly French) poststructuralist targets in the late twentieth century. In these academic spats, there are repeated attempts to unveil the other or expose pretensions to knowledge as charlatanism.[9] Parody and the act of unveiling or exposure are central to Sedgwick's understanding of paranoid practice. But, if science is under ubiquitous pressure to perform and clearly enmeshed within the structures of late capitalism, what use is repeatedly revealing this, as if it is hidden, when it is all too visible? Similarly, when scientists expose limited understanding of quantum mechanics among poststructuralist philosophers, who, in this cynical, paranoid world, is really surprised or cares? Perhaps more importantly though, parodic acts of exposure may just serve to reinforce the paranoia and cynicism that they intend to target. The Indian scholar, Shiv Visvanathan, finds appeal in Alan Sokal's hoax as joke or trickster performance, but wonders if 'the same joke sounds poor when seen from the life-world of a tribal, a marginal or a slum'. As he goes on to say, 'jokes from the system are not equivalent to laughter in a life-world' (Visvanathan 2005, 83).

Exposing hoaxes is a key dramaturgical weapon of epistemological warfare and was a strategy of Donald Trump's successful presidential campaign in 2016. His supposed act of exposure was to dismiss the scientific evidence in favour of anthropogenic climate change as a hoax perpetrated by a variety of disparate forces including Barack Obama and the Chinese government. In a 2015 speech at a rally in South Carolina, Trump claimed 'Obama's talking about all of this with the global warming, da la da [gestures] it's a hoax, a lot of it's a hoax, it's a money-making industry ok' (C-SPAN 2015). When US presidents are proposing 'alternative facts' of such audacity and magnitude, it is questionable how much critical energy should be devoted to unveiling the social construction of scientific facts.[10] This book therefore searches for creative alternatives to epistemological warfare through aesthetic vocabularies that go beyond or at least nuance the established tropes of science in performance: the parodic unveiling of the charlatan, myth-busting and, on the other side, dialectical critique or the tragedies of scientific hubris.

In a short chapter responding to a call for creative approaches to bringing nonhumans into democracy, Isabelle Stengers imagined an alternative to a politics of critique which she articulated as a 'cosmopolitical proposal'.[11] The cosmopolitics that she briefly explains in this chapter and at much greater length in her books is a practical politics that has little to do with a unifying theory or model of global citizenship, as in Kantian cosmopolitanism. Rather, Stengers' cosmopolitics is not so much a political theory as a stance or attitude which she identifies with the figure of *The Idiot* from Dostoevsky's 1869 novel. The idiot does not question knowledge per se but 'demands that we slow down, that we don't consider ourselves authorized to believe we possess the meaning of what we know' (2005, 995). The stance of the idiot enables us to ask ourselves if there might be anything more important than what we know.

Part of the cause of epistemological warfare, for Stengers, is a tendency towards polemic in claims for truth. It is perhaps easy to understand this polemic in terms of a series of oppositions: humanities vs. science, myth vs. science, relativism vs. positivism, nature vs. culture. However, Stengers outlines a more nuanced version of the knowledge rupture that, in an unproblematically reconciled format, appears in Stephen Hawking's line at the opening of this chapter:

> We live in a universe governed by rational laws that we can discover and understand. Look up at the stars and not down at your feet. Try to make sense of what you see and wonder about what makes the universe exist. Be curious.

The imperatives at the end of the statement are direct calls for engagement: demands for affect rooted in wonder and curiosity. In the Paralympic Opening Ceremony they seem to be a demand for a leap of faith embodied by Miranda/Miles-Wildin's flight across the stadium. This emotional realm and its sense of movement though have historically been at odds with science's claims for a privileged relationship with rationality and sets of discoverable laws contained in the opening sentence.

If there is one polemic that creates a break between practices of science and practices of non-science, there is another polemic within science about how it distinguishes itself from the non-scientific. Stengers has traced this doubled polemic back to the way sciences have differentiated themselves from (sophistic) rhetoric, on the one hand, and from myth, on the other:

> The sciences follow a narrow path, ever on the defensive against the powers of the imagination, which are satisfied with explanations and significations forged without constraint, and against the powers of rhetoric, which are satisfied with the ambiguity of language and the pretenses of proof. (Stengers 2010, 2)

Galileo's battle with the church, for instance, is often given as an example of a struggle of science against the power of myth. However, this struggle is doubled by more questions. Is scientific practice the act of looking up at the stars, so scientific on the basis of its empirical and therefore human methodological approach? Or is science the (passionate) belief or faith in the possibility of coherent understanding of what makes the universe exist that goes beyond the possession of this knowledge by humans?

Stengers illustrates this second polemic through an account of an argument between two physicists, Max Planck and Ernst Mach, at the start of the twentieth century. For Planck, the truth claims of physics went beyond methodology, they rested in a *belief* 'in the possibility of achieving a unified concept of the physical world' rather than Mach's more 'pragmatic' conception in which such truths were tied to the perspective of the human observer (Stengers 2010, 6). Stengers draws attention to Planck's position as one of belief or faith that is erroneously equated with rationality. Planck's position is one of 'scandalous creativity' that is opposed to Mach's critical rationality (Stengers 2010, 8). Planck's faith, importantly, does not devalue the truth claims of physics or science more broadly. It is not reducible and cannot be explained away in terms of culture or society. According to Stengers, in what might be called a scientific performative, Planck 'cries out… the *fact*' of scientific truth (Stengers 2010, 7, italics in original). The celebration of the Large Hadron Collider I discussed at the start of this chapter seems to me to echo forth Planck's cry. One political problem arises when science is simply explained away as just another truth-claiming practice equivalent to any other, as President Trump seems to. Another arises when sciences assume or are given exclusive claims to rationality. This both devalues science as a set of *passionate* practices – acts of faith – and tends to prematurely foreclose on the truth claims of other practices. Stengers has commented elsewhere on the sciences that it is 'the role of guardians of rationality that has captivated them and put them at the service of power, both state and capitalist power' (2011a, 375). Within modern science, Planck's and Hawking's passion is under both performance pressure and the threat of reduction by critical rationality. This passion of and for science is often elided in characterisations of scientists as coldly rational and sometimes disavowed by scientists themselves. Epistemological warfare between sciences and humanities or science and art gives in to a definition of truth which requires its transcendence in a politics of *either* socially constructed *or* scientifically true. Stengers calls for practices that resist making such a choice. Rather than pointing out 'the relativity of truth', she asks us to 'affirm the truth of the relative', what she calls 'the humor of truth', an idea I will return to directly in the next chapter (Stengers 2010, 11).

This is why this book constitutes a search for moments in theatre where critique is suspended, where science appears as passionate practice, where reductions of various kinds are temporarily resisted. There is a danger that such an endeavour appears perverse or even wilfully stupid when knowledge of all kinds faces pressing needs and there is widespread thirst for critiques of injustice or inequality. It is easy to condemn scientists as frivolous when they stop their work to collaborate with dancers, or question what dermatologists have to say to grime MCs. Similarly, such collaborative practices may be seen as a selling out by activists or radical artists, a departure from the development of critical accounts of epistemological hegemony. I own up to a certain degree of determined idiocy in turning to theatre in the face of such challenges.

Inspired by Stengers' cosmopolitical proposal, this book explores the possibility of faith in a politics of idiocy attempting to identify potential cosmopolitical practices within theatrical performances. Such a politics relies on Stengers' and indeed Dostoevsky's conception of idiocy, which does not reflect a lack of insight or intelligence. It retains its pejorative connotations but, in the face of these connotations, accepts the risk of ignorance or not-knowing. Dostoevsky's 'idiot' Prince Myshkin is a Christ-like figure who, in his journeys through the experimental novel, offers a diagnosis of the social conditions of mid-nineteenth-century Russia.[12] Rather than propose theatre or indeed theatre studies as alternative modes of knowledge production to either science or social science, this book has the more modest aim of proposing theatre and the theatrical as practices of creative ignorance: stances from which slower more careful modes of knowledge-making might be considered. Within the politics of idiocy and the practices of repair, 'mere-ness' becomes an ethical principle. Such politics find in theatricality a stance that is unashamedly 'merely aesthetic' and 'merely reformist' and consequently merely pleasurable and merely ameliorative. But as Sedgwick asks, 'what makes pleasure and amelioration so "mere"?' (Sedgwick 2003, 144). I argue that in its mereness, theatre might contribute to what Stengers calls a 'diagnosis of becoming' that rather than expose or critique science can 'risk assuming an inventive position that brings into existence, and makes perceptible, the passions and actions associated with the becoming it evokes' (Stengers 2010, 12).

Making sense of science in the theatre

Science studies from a variety of philosophical and disciplinary perspectives have not missed the potential of theatre or drama as metaphor or theatricality as a concept for making sense of science. In his study of the US National Academy of Science, sociologist Stephen Hilgartner (2000, 7) has analysed science advice to policy makers as 'public drama'. He employs a theatrical metaphor for examining how science advisors 'present themselves to audiences' and acquire credibility. The philosopher of science Robert Crease uses theatre as an analogy in understanding experimentations within science. The theatrical analogy for Crease is:

> [a]n organization tool to hold together the various hermeneutical dimensions… [and] as a 'physiognomic' tool, in allowing us to reorganize our perception of science and letting us see and talk about aspects of experimentation that have been overlooked in the mythical view – to *constitute*, as it were, a new perception of science itself. (1993, 96)

In *Pasteurization of France*, Latour makes a similar move trying to understand the emergence of French science through a semiotic analysis of its actors and narratives (1988, 11). He turns to a theatrical analysis as a way of bringing a wider variety of actors onstage than histories that focus on Pasteur in their retrospectively constructed account. Thinking about the emergence of the germ theory of disease as theatre enables Latour to pay particular attention to the way scientific proof is staged and how its audiences are positioned. Latour's particular dramaturgical approach is notable for its allocation of roles to the microbes as well as a variety of human actors.

All of these accounts perhaps overstate the utility of theatre in revealing something that is hidden or overlooked. Theatre, after all, does not make visible things that are really there but cannot be (easily) perceived; this is what science does. Rather, theatre tends to make perceptible things or attributes of things that are not really there alongside, or at the same time, as things that are. However, what I think they share with my argument is a recognition of the particular way that theatre constitutes a condition in which contradictory modes of being might co-exist. Crease refers to this condition as 'antinomic' and talks about theatrical performances as 'at once unique events in the world, yet [that] at the same time reveal something *about* the world' (1993, 98). This condition is appealing to Crease because he also regards science as antinomic. However, unlike science, theatre is much more variable and ambivalent about the extent to which its practitioners seek to hide or disavow the ambiguity of its condition. Even theatre's more anti-theatrical practitioners tend to acknowledge or even rely on the limitations to theatre's revelatory potential. Theatre is important to cosmopolitics or the politics of idiocy because of the way, as a stance, it offers ways of resisting dramaturgies of exposure, revelation and unveiling. It is also important to cosmopolitics in its attention to and reliance on particular practices for tracing associations and assembling collectives on stages and in, through or with audiences.

For Josette Féral and Ronald Bermingham (2002, 97), theatre is constituted through processes of theatricality that produce a distinct 'spatial cleft' in daily life. Theatricality 'has to do with a "gaze" that postulates and creates a distinct, virtual space belonging to the other, from which fiction can emerge'. This cleft can be created either by the spectator, or invariably in theatre and other kinds of theatrical performance through some degree of complicity between the spectator and actor. Theatricality thus conceived is not co-extensive with theatre although theatre is of course exemplary as a theatrical space. Theatricality, thought of as a process of differentiation from everyday life, is a process by which unreality is constructed. Theatricalising science involves either the construction of its practices, its methods or its objects as unreal through performance, spectating or a combination of the two. By understanding more about how theatricality makes science unreal, certain insights may also be gained both about what science does to theatre's untruth claims and indeed (by a process of reverse engineering) how science might or might not resist threats to its truth claims posed by more or less theatrical performance pressures.

The central role of the spectator or audience in constituting theatrical engagement is an important theme throughout this book. Public understanding and public engagement with science in their inception as movements have tended to caricature or make generalised assumptions about their publics. Historians of science communication Jane Gregory and Steve Miller have pointed out that within the public understanding movement and its predecessors 'the public and their understanding are black-boxed to alleviate the need to decipher them' (Gregory and Miller 2000, 95). A black box, in this sense, is an analogy for a closed or settled concept or technology whose internal workings, as opposed to what goes in or comes out, are no longer a matter of concern.[13] So while much of this book documents what happens on stages, it also pays considerable attention to the ways in which audiences are assembled, composed, positioned and affected.

Scholars in science studies have not been shy in their use of theatricality and the theatrical metaphor. Bruno Latour has even turned theatre producer in his involvement with the 2014 play *Gaia Global Circus* by Pierre Daubigny (2018). However, such scholars have been largely neglectful of wider research in theatre studies. This does mean that such texts tend to easily conflate concepts such as theatre, drama, performance and theatricality. It also means that the concept of theatre or drama in use as metaphor or analogy reflects a number of unstated assumptions about theatre practice. Greater attention to theatre studies might shape understanding across a number of disciplines of how science performances have been co-produced in relation to performance practices, theatre as social institution, theatricality as process and drama as structure or medium. Turns to performance, theatre and theatricality by a number of scientists and science communicators, not to mention philosophers and sociologists of science, may, on the one hand, reflect the performance pressure outlined at the start of this chapter. On the other, theatricality's tendency to differentiate space from the everyday and constitute it as unreal, perhaps reflects a rather different trajectory that has less to do with concerns about whether an object is constructed, but how. It is this potential of theatre and theatre studies as ways of articulating processes of construction and co-production that is recognised and exemplified in a diverse range of practices from the first decades of the twenty-first century to which this book hopes to contribute.

It is difficult to make specific quantitative claims about the extent to which particular forms of cultural production such as theatre, even narrowly conceived, thematise or otherwise engage with such a capacious and contested set of practices as the natural sciences. Nevertheless, a number of commentators have argued that 'science plays' have proliferated in the English language theatre since the 1990s. Kirsten Shepherd-Barr takes what she regards as a 'recent explosion of plays about science' on professional stages in the UK, USA and Europe as part of the rationale for her attempt to provide a historical context for the phenomenon and to work towards a taxonomy (2006, 2). Eva-Sabine Zehelein, in a study that focusses on the USA and the UK 1990–2007, goes further in arguing that, within her period of study, 'more theater and performance projects have integrated the natural sciences than ever before' (2009, 3). The relationship between science and theatre, however, has an extensive history with persistent tropes as Shepherd-Barr and Zehelein helpfully show.

This relationship cannot be understood exclusively in terms of the extent to which playwrights and producers on or off Broadway or London's West End stage science as narrative or even use scientific concepts to inspire new modes of dramaturgy. However, such particular contexts can be indicative of wider cultural currents and show how theatrical practices give a particular and valuable perspective on science and its societies. Indications of this potential can be seen within a number of theatre histories that unpack the co-production of evolution by scientists and theatre-makers (Goodall 2002; Shepherd-Barr 2015; Wolff 2014). Similarly, theatre and performance studies have also offered important perspectives on the co-production of the human body by scientists, doctors, patients and other performers at the interface between theatre and medicine (Mermikides and Bouchard 2016), theatre and cognitive science (Blair and Cook 2016), or theatre and multimedia technologies (Parker-Starbuck 2011).

Such accounts work across and between representations of figures in a range of scenarios. Histories of science plays have characteristically focussed on the figure of the scientist. Science plays in these cases are often constituted dramatically by interplay between what scientists do inside science and what they do outside it. Scientific or non-scientific scenarios are literally spatialised in the scenography, or the drama emerges from the interdependence of scientists and non-scientists within a shared scenario. In this way drama is constituted by the tension between the scientist's objective perspective on nature and the scientist's position as subject in and of drama. Through this relationship between figure and scenario, such theatre explores the emergent co-production of nature and culture. Within medical dramaturgies a significant political shift is marked by moves from the doctor as central figure, for instance in Shaw's *The Doctor's Dilemma* (1987a) or Ibsen's *An Enemy of the People* (2008), to the patient, for instance in Margaret Edson's *Wit* (2000). Such shifts in subject position reflect the emergence of and then reaction against the 'medical gaze' and its objectification of the human body as influentially theorised by Michel Foucault (2003) in *The Birth of the Clinic*.[14]

A further shift in subjectivity emerges in what might be called biosocial theatre where the relationships constituting biosociality discussed above are enacted or re-enacted within the theatrical context. This may follow dramaturgical conventions associated with documentary forms reflected in some of the practices discussed in Chapters 4 and 5 of this book or indeed 'post-dramatic' or live art approaches in which the patient or health professional is recounting their own experience.[15]

However, theatre and performance studies have also offered up figures and scenarios that cut across established notions of subject and object or nature and culture in their exploration of the co-production of theatre and science. Sue-Ellen Case (2007, 2) explores 'strategies of representation' across science and performance through the mobile and versatile figure of the avatar as it operates within virtual space. The avatar's traversing of virtuality enables her to connect alchemist to scientist, Mme Blavatsky to Sun Ra or Lara Croft in ways that challenge and connect disparate fields. Jennifer Parker-Starbuck (2011, 6) has used Donna Haraway's figure of the cyborg to track the interweaving of the human body and technology across contemporary media. The rapidly expanding field of environmental or eco-critical theatre and performance studies has explored scenarios of environmental catastrophe following established humanist critiques of the despoliation of nature, but also through a number of innovative nature-culture scenarios.[16]

Rather than offer up a single exemplary figure or scenario though, this book attempts to multiply the possible vocabulary and perhaps more importantly also to draw attention to the way engagement and co-production are constituted through *theatre* practice. I share with Parker-Starbuck the sense of theatre as 'a necessary term, a seeing place but also a meeting place within which we can confront and question others and ourselves' (2011, 7). This is a book about practice as a source of repair.[17] Such practices take place in and through time and space. A key aim of the book is to assemble descriptions of these practices in their temporal and spatial dimensions. It is not a work of science, philosophy or sociology; neither is it an act of epistemological warfare. As such it is not trying to sketch out a new interdisciplinary field, a mode or model of interdisciplinarity. However, I do hope to offer a certain political perspective on particular discourses of interdisciplinarity.

A vocabulary of interdisciplinarity has grown up often explicitly in response to the epistemological warfare discussed above. Art-sci, sciart, bioart, molecular aesthetics, all reflect different discourses on sets of overlapping practices through neologisms or paired concepts with distinct aesthetic and political implications.[18] There are also an increasing number of organisations and institutions that support interdisciplinary practices that cross over between arts and sciences, notably ZKM in Germany, Ars Electronica Centre and Festival in Austria, Le Laboratoire in France, the network of Science Galleries, Arts Catalyst in the UK and SymbioticA in Australia. In both theory and practice, visual artists, including some working in performance media, have been more closely involved in these developments than theatre-makers. However, a parallel set of terms has also emerged that might include science play, science theatre, science-in-theatre and affective performance.[19]

Some theatre-makers and theatre companies have developed temporary or longstanding commitments to particular ways of working with science and scientists. A number of theatres in the UK including the Theatre Royal Plymouth, Soho Theatre and the Summerhall venue at the Edinburgh Festival Fringe have staged programmes of work connected by a particular conception of science. The Ensemble Studio Theatre in New York has produced an extensive repertoire of plays funded through their partnership with the Sloan Foundation. Certain touring companies have developed extensive track records and approaches to making theatre on scientific themes and in collaboration with scientific experts: for example, Y Touring Theatre Company's Theatre of Debate (discussed in Chapter 4) and Curious Directive in the UK or Jean-François Peyret in France. Theatrescience formed by Jeff Teare and Rebecca Gould developed an original process which incorporated young people alongside scientists and playwrights in theatre projects around the UK, in India and between the two countries. Activist arts organisations such as Platform and Critical Art Ensemble have created distinctive actions combining science, politics and experimental performance in many different locations. Museums and interactive science centres, from the Exploratorium in San Francisco to Unizul in South Africa, have played a major role in developing particular approaches to science in performance. These practices are clearly important. However, it is not my intention to identify or privilege a particular approach to theatre-making. Neither do I want to delineate a novel interdisciplinary aesthetic or taxonomy of performance. Rather, I am interested in how a wide range of theatre and performance practices, including some unexpected or challenging examples, might offer cosmopolitical insight or make links between different ways of knowing.

Such a project does of course have implications for interdisciplinarity as a discourse and for practices that make particular claims to interdisciplinarity. Andrew Barry *et al.* (2008) have identified three 'logics of interdisciplinarity' in a study that included, but was not restricted to, examining art-science collaboration through the discourse of researchers and policy-makers. They identify logics of accountability and innovation that might be associated directly with the performance pressure discussed at the start of the chapter. So the rationale for interdisciplinary collaboration between scientists and theatre-makers might be to foster greater democratic accountability in response to the crisis of trust discussed above, or such collaboration may constitute a mode of technological, social or cultural innovation within the framework of the knowledge economy. Born and Barry (2010) particularly associate science and art funding programmes in the UK with these logics and comment critically on how these logics 'interpellate' publics filling the empty but fixed space – the black box referred to by Gregory and Miller. Science as a fixed body of knowledge is communicated to or understood by a preformed notion of the public

The third logic they identify is 'an orientation towards effecting ontological change' through rethinking 'both the object(s) of research and the relations between research subjects and objects' (Barry *et al.* 2008, 25). As they point out, all three logics may often be in play in relation to particular practices. They are 'entangled', but 'irreducible' to each other. I find the logics of accountability and innovation more obviously identifiable within policy or institutional discourse than in embodied practice. Theatre of Debate's practices, for instance, have been rationalised in funding applications and evaluations in terms of democratic accountability and science education. However, their particular theatrical forms of debate, discussion and dramaturgy resist evaluation in simple terms of increased understanding or even as democratic participation. Born and Barry (2010, 116), following Hannah Arendt, identify the ontological logic of interdisciplinarity particularly with the performing arts as a 'preferred idiom' insofar as it is produced at the same moment and through the assembly of its public. Theatrical performance in particular is a peculiarly temporary, contingent, artificial and uncertain mode of existence.

Debates about liveness, theatre and mediatisation have shown the way theatre has a peculiar ontological status (Auslander 2008; Meyer-Dinkgräfe 2015; Phelan 1993). It is distinguished by being indistinct. It is often articulated in paradox and contradiction and through doubled formulas. The most widely used of these perhaps come from Victor Turner's and Richard Schechner's ideas: 'twice-behaved behaviour'; 'not... not-not'; and liminality that is not just spatially between fields but betwixt *and* between in the passage between time-spaces (Schechner and Turner

1985; Turner 1970). I am not therefore arguing for theatre's potential as linked to its distinct spatiality as in a liminal or interdisciplinary *space* or proposing that drama transforms through its narrative structuring of *time*. Rather, theatre diagnoses becoming through its combined, ambiguous and often contradictory operations in (and out of?) *both time and space*, and I will return to exploring the time-spaces of science in performance in the last two chapters of the book. Theatre copes with contradiction through paradox, juxtaposition, counter-position and other strategies for holding things together. As Baz Kershaw has commented in his work on *Theatre Ecology*, paradox is 'crucial' to theatre. It emerges from convergence of apparently contradictory ideas but resists resolution or integration. Paradox, for Kershaw, 'includes but travels beyond the binaries of body and mind, female and male, nature and culture and all the rest that dog ecological wisdom' (2007, 23). Paradox is therefore central to how theatre makes sense of science and central to cosmopolitics as well as other forms of political ecology.

In the remaining chapters of this book I examine theatrical practices of sense-making that include: speculation, comedy, choreography, debate, translation. These concepts describe movement between time-spaces rather than stake out space between. They invite sense-making practices that recognise science's passions at the same time as they articulate feelings from emergent levels of sensory perception. In the next chapter, I will explore how theatre enables laughter at science as a particular mode of engagement that acknowledges what Stengers calls 'the humor of truth' (2010, 11). In this conception theatre does not produce engagement; it is a particular and peculiar mode of engagement. While theatre audiences may form a proxy for a public, a crowd, a group of consumers or a nation, they are not quite any of these collectives. Rather, the theatrical events described in this book constitute collectives in mobile ways, tracing emergent associations and solidarities articulated through various modes of sensory perception.

Notes

1 Global television viewing figures are contested. Channel 4 achieved a UK audience of 11.2 million.

2 The Paralympics, including its 2012 opening ceremony, is a site of paradoxical performances in which science is complicit in uniting and dividing its polity. The sporting spectacle celebrates human diversity and, in its multiple quasi-scientific categorisations of disability, relies on a complex institutional apparatus of discrimination between performances. The work of the International Paralympic Committee is underpinned by and also provides a programme for a particular strand of sports science.

3 The model of the universe known as the Big Bang is normally attributed to the Belgian astronomer Georges Lemaître and a paper he published in 1927. However, the term 'Big Bang' was originally coined by the British astronomer Fred Hoyle in 1949 when referring rather dismissively to Lemaître's theory (Lovell 2001; van den Heuvel 2016). The umbrella was of course not invented in Britain in the nineteenth century, but British manufacturers pioneered the use of steel ribs in their mass production in this period (Fox Umbrellas Ltd 2014). The history of the orrery is also somewhat contested; however, George Graham is acknowledged as an early maker (Westfall 1995).
4 Technoscience as a concept is normally traced back to the work of Gaston Bachelard (1984), but has been popularised and extensively discussed since the work of Donna Haraway in the USA and Bruno Latour in France (Haraway 1997; Latour 1987).
5 There is extensive literature offering historical and political analysis of public engagements with science within the pages of academic journals such as *Science Communication* or *Public Understanding of Science*. Jane Gregory and Steve Miller's *Science in Public* (2000) offers a comprehensive historical account of the US/UK context up to 2000.
6 The health research charity the Wellcome Trust (founded in 1936), the parent organisation of Wellcome Collection, now clearly forms part of the institutional (co-productive) infrastructure of science in the UK alongside the more venerable bodies of the Royal Society (1660) and Royal Institution (1799). It was established as the philanthropic legacy of the pharmaceutical entrepreneur Sir Henry Wellcome, but its influence on science policy has grown significantly since the diversification of the Trust's share portfolio in the 1980s and 1990s orchestrated by the City financier Sir Roger Gibbs. Its spending on (primarily biomedical) research grew from £50 million in 1989 to £1 billion 30 years later. The Wellcome Trust was a key player in the emergence and evolution of public understanding of science and public engagement with science during this period and has devoted a significant proportion of its funding towards practices of engagement and education in diverse forms, including the establishment of the Wellcome Collection in 2007.
7 Henry Wellcome made his fortune in collaboration with Silas Burroughs from the emergent pharmaceutical industry pioneering technologies of standardising drug doses through compacted pills (Launer 2017). Gates pioneered software for personal computers and Howard Hughes had a range of business interests notably in aviation and film production.
8 There is a strand of contemporary science in performance, ranging from Richard Dawkins' brand of science communication to the musical stand-up of Tim Minchin and the science rap of Baba Brinkman that targets particular mystical (or pseudo-scientific) practices for satire, perhaps because they are relatively soft targets or because they are seen as symptomatic of anti-scientific malaise or post-truth society.
9 Stefan Collini's introduction to a 2008 edition of Snow's *The Two Cultures* provides an explanation of the debate with Leavis and also its relationship with the earlier Huxley vs. Arnold debate (Snow 2008). Alan Sokal performed

the exemplary 'unveiling' of poststructuralism in his hoax article published in the pages of *Social Text* (Sokal 1996). In a dramatic act of exposure after publication of an article that referred to ideas of 'quantum gravity' within a critique of scientific epistemology, he revealed the parody and, or so he intended, showed the world that the postmodern emperor had no clothes. The 'Sokal Affair' is well documented in a variety of texts, notably those by Sokal himself (Sokal 1998). It remains to be seen what effects this dramatic act had within or without the ivory towers that physicists and philosophers often co-inhabit.

10 Latour (2004a) makes this point in an article in which he derives his own version of reparative reading calling for a stance in which adding to matters of fact replaces the impetus to subtract and objects are assembled and reassembled rather than deconstructed.

11 The chapter appeared within a large publication connected to a 2005 exhibition entitled *Making Things Public* curated by Bruno Latour and Peter Weibel at the ZKM in Karlsruhe.

12 Myshkin's idiocy in his own eyes and in the eyes of other characters combines naivety with insight, derived partly from experiences of illness, and openness to the experiences of those marginalised in society. This is exemplified in his narration of his relationship with some children and a young woman while in Switzerland (Dostoevsky 1869, 84–9).

13 Richard Whitley used this term influentially to critique the 'black-boxing' of science by sociologists (Pinch 1992). Critiques of black-boxed ideas of the public, albeit using different geometric analogies notably that of the sphere, might be found in a strand of political theory from Habermas (1992) through to Nancy Fraser (1990) and Michael Warner (2005). Scholarship exploring the construction of the theatre audience may also be understood as opening a black box (Bennett 1997; Freshwater 2009; Sedgman 2018; Sedgman and McGrath 2016).

14 Much fuller accounts of this trajectory can be found in the literature I have referred to (Mermikides and Bouchard 2016; Shepherd-Barr 2006).

15 For example, see extensive work by Clod Ensemble, Bobby Baker, Bryony Kimmings, James Leadbitter, Brian Lobel, Peggy Shaw and Martin O'Brien (Barrett and Baker 2007; Kimmings *et al.* 2018; Lobel 2008; Mermikides and Bouchard 2016; Parry 2017).

16 There are too many texts to cite all of them. Practice is well documented through the impressive Ashden Directory and discussed in a number of critical sources (Chaudhuri 1994; Giannachi and Stewart 2005; Kershaw 2007; Lavery and Finburgh 2015).

17 Understanding theatre and performance practice in terms of repair is a central idea in published reflections on the work of US company Goat Island (Bottoms 2007). These reflections, referred to in texts by Stephen Bottoms and Matthew Goulish, resonate with conclusions I reach in this book, for instance in relation to theatrical practices of repair, slowing down and theatre's potential contribution to political ecology.

18 There is a growing literature that documents and theorises collaborative or interdisciplinary practices between the sciences and arts (Arends and

Thackara 2003; Ede 2004; Edwards 2009; Kac *et al.* 2009; Kemp 2006; Miller 2014; Weibel *et al.* 2013).

19 'Science-in-theatre' was delineated as a genre by Nobel prizewinning scientist turned playwright Carl Djerassi (2002). The science play has been documented and discussed by a number of theatre historians (Bartleet 2013; Kazzazi 2017; Shepherd-Barr 2006; Zehelein 2009). Affective performance emerges as concept from the intersection of research in applied theatre and cognitive sciences (Shaughnessy 2012).

2

Laughing at science in the theatre: Gob Squad, a funny robot and dancing scientists

In a fleeting moment at a rehearsal for *My Square Lady*, in a large space backstage at a Berlin opera house, Myon, the humanoid robot and ostensibly central character of the piece, turned its head and focussed its gaze on me. I briefly appeared on the screen overhead showing Myon's periodically changing and unpredictable point of view. I laughed quietly to or maybe at myself. In the notes I made at the time about this rehearsal session, I also remarked that one of the production team giggled and repeatedly went 'ahh' every time Myon's viewpoint changed. It sounded to me like the kind of noise fond adults make when a child does something cute or endearing. This made me laugh too.

The rehearsal days I attended in May 2015 were near the end of a two-year collaborative process involving artists from the German/UK Gob Squad art collective, the Komische Oper in Berlin and a group of researchers from the Neurorobotics Research Laboratory at Beuth University of Applied Sciences. I had been attracted to the project in the course of research for this book partly just because it involved a long period of collaboration between theatre-makers and scientists. However, there was also something about the apparent zaniness of the project that made it stand out and drew me to Berlin. Both its title and the idea behind it had made me laugh even before I started work on it. The name of the project was of course a play on the title of Lerner and Loewe's musical *My Fair Lady*. The 1956 Broadway show was an adaptation of G. B. Shaw's 1914 play *Pygmalion*, itself inspired by the classical story of Pygmalion and Galatea.[1] Both Shaw's play and the musical are set in Edwardian London, where the arrogant linguist Professor Higgins bets that he can teach the flower-girl Eliza Doolittle to pass for a duchess at a high society event. In an experimental move, Gob Squad would cast Myon the robot as Eliza. Instead of the working class girl passing for an aristocrat in Edwardian England, the posthuman robot would attempt to pass for a human performer on the main stage of the prestigious opera house in central Berlin.

Gob Squad member Sarah Thom explained to me how the project actually started off with this joke, emerging out of hilarity at the collective's annual creative retreat:

> [w]e became more and more interested in the Pygmalion thing and the opera and then we realised *My Fair Lady*, and then as a joke we said why don't we call the piece *My Square Lady* and then we thought that's hilarious and we'll use that as a working title.

Gob Squad's experimental creative approach during the retreat involved members of the art collective proposing apparently unrelated personal interests, looking for connections between them and then with external opportunities. In this case, interests in robotics, science fiction and artificial intelligence came together with Ovid's *Pygmalion* story, a desire to explore relationships with objects and an urge to make opera. This unlikely combination articulated with a funding opportunity presented by the German government's *Doppelpass* fund for cooperation between smaller experimental touring theatre companies and larger established building-based theatre institutions.[2] The funding meant that Gob Squad could collaborate over the two-year period with a group of opera singers, a production team and other groups from the Komische Oper and researchers from the Neurorobotics Institute.

Over the course of the process, Gob Squad became embedded in both the opera house and the robotics laboratory. They had many conversations with backstage staff and performers at the opera as well as with researchers in the lab. At various points they took the research robot Myon into the theatre to join in these conversations and, using its recording equipment, documented aspects of the process. This evolved further in a variety of creative interactions between the robot and performers at the opera. Members of the robotics research team started visiting the opera house and, along with Myon, ultimately found themselves in rehearsals and then on stage.

The collaboration culminated in a short series of performances on the main stage of the opera house in summer 2015. As I will discuss in more detail below, these were celebratory events involving a large eclectic cast and all the theatrical infrastructure of a large state-funded opera. The singers from the opera house performed a selection of popular arias and other songs from classical and pop repertoires. The opera house's children's choir and adult chorus appeared in robot costume boosting the volume and spectacle. Myon the robot appeared 'live' (in the flesh/plastic) and its point of view on the process was edited into video clips projected onto a screen on stage. In conversational interludes, Gob Squad members questioned opera singers and scientists about their work and their personal life. And in several show-stopping numbers, there was dancing, singing, a scientist flying across the stage, and a poignant moment when the whole cast theatrically dropped dead. The performance mixed profound philosophical questions with razzmatazz and a lot of fun and laughter.

It would be easy to regard as frivolous such a combination of comic opera and the pop culture-infused anarchic aesthetic that Gob Squad has a reputation for. Perhaps for this reason, my first drafts of the start of this chapter were more serious attempts to contextualise this project in relation to understandings of robotic science and its publics. I will get onto this. However, I gradually realised that it was the laughter and sense of fun that I needed to take more seriously. This is what constituted the engagement between scientists, artists and publics; so I returned to the sketchily recorded moments above as an appropriate point of departure for reflecting on a highly serious and systematic attempt to have fun with scientists and make people laugh fondly at science. I want to argue that laughing at science may in fact be an important mode of public engagement and political practice of repair. I identify Eliza Doolittle and indeed Professor Higgins, entangled as they have been with the developments in artificial intelligence and computer sciences, as key figures of fun and therefore enduring agents of theatrical repair.

Learning to laugh at science is surprisingly hard and holds significant political potential. Many of those who defend science are understandably concerned that subjecting science to comedy through parody or satire may be to undermine it, equating laughter with derision or the loss of trust discussed in the previous chapter. However, laughing need not be mocking or derisive. Rather, in this chapter, I argue that laughing at science may offer a mode of engagement that resists making a choice between invidious alternatives in which scientific practices are equated either with unique access to truth via their objectivity and superior rationality, or are understood as entirely the function of social relations. Isabelle Stengers (2000a, 52) has argued that to laugh at science is to resist being 'easily impressed', not by taking a superior, critical position but by feeling able to complicate problems without offering premature solutions.

Laughter can avoid reductionism including the reduction of diverse scientific practices to a singular ideology of rationality, objectivity or truth. Stengers calls this 'the laughter of humor' and identifies it as a particular feminist position, not proposing an alternative science, but rather offering a challenge to the dominant ways in which science and society are co-produced. Such laughter is 'comprehending and appreciative without expecting to find a secure position. It is able to disagree without being awe-stricken or trying to be awe-inspiring' (Stengers 2000a, 52).

So what was funny about *My Square Lady* and what political potential might that funniness hold? What made me laugh in that rehearsal? My laughter no doubt reflected some anxiety in my relationship with the technology. It was perhaps a response of surprise at the unexpected. Maybe there was some narcissistic joy at somehow becoming part of the phenomenon I was studying. The French philosopher, Henri Bergson (2010, 20), identified people pretending to be things as a starting point for his theory of laughter famously formulated as 'something mechanical encrusted on the living'. Perhaps my laughter was prompted by what appeared here to be the reverse operation – something living encrusted on the mechanical. However, this account seems to suggest a relatively simple operation in distinguishing the mechanical from the living. Part of what seemed to be going on here was a confusion of the two. Besides, I am not sure if I was laughing at Myon, myself or my reaction to Myon. Similarly, when the member of the production team giggled and I laughed at the laugh, what was going on here? At this point we became somehow aggregated as an audience or laughing public: laughing at Myon, laughing at or with each other, laughing at ourselves. At this point easy distinctions between subjects and objects seem to break down. Did Myon make me laugh? Did I make myself laugh? Did my laughter at Myon or the member of the production team subjectify or objectify them or me?

In their introduction to a special issue of *Critical Inquiry*, Sianne Ngai and Lauren Berlant have identified the difficulties both in theorising comedy in formulaic terms and conceiving of laughter as a simple outward expression of comprehensible internal feelings. Comedy for them is 'epistemologically troubling, drawing insecure boundaries as though it were possible to secure confidence about object ontology or the value of an "us" versus all its others' (Berlant and Ngai 2016, 235). Laughter has the potential both to trouble knowledge, what we know, and trouble the perspectives from which we might know things, so who we are and how we know. Laughter as public engagement holds political potential because it both seems to arise from subjective feelings about objects but also renders us aware of ourselves as objects. It may be possible to establish clearly how to provoke laughter, as Georges Bataille suggested, but its causes may be unknowable or, as Bataille then puts it, its cause may indeed be '*the unknowable*':

> We laugh, in short, in passing very abruptly, all of a sudden, from a world in which everything is firmly qualified, in which everything is given as stable within a generally stable order, into a world in which our assurance is overwhelmed, in which we perceive that this assurance was deceptive. (Bataille and Michelson 1986, 90)

Public laughter, including laughter in the theatre, constitutes a political mode of engagement with science because of the way that it marks and draws attention to what is not known and, perhaps in some sense, what might be unknowable, at the same time as putting into question the position of those laughing. It is a necessary and appropriate response to areas of science that are the most politically troubling and that are the main sources of inspiration for the practices discussed in this book: climate sciences, biomedicine and, in this chapter, artificial intelligence.[3]

If computer sciences and the technologies of artificial intelligence trouble us, it is partly because it can feel as if they are driving social change at an increasingly rapid pace. However, as I have explained in the previous chapter, my perspective in this book reflects on and through the notion of co-production. So I am trying to resist the idea of a simple technological determinism. Rather, I am trying to explore how computing technologies and social relationships form each other. As Sherry Turkle (2012) has pointed out, there is a persistent tendency for humans to attach and become intimate with new technologies. At the same time such technologies exert significant power and control over us. The technologies and our relationships with them become very quickly 'naturalised'; that is to say, it becomes very hard to imagine our way out of the relationship or to reimagine the technology or our mode of interacting with it. New technologies emerge with particular social relations embedded into them and in a 'reverse feedback loop' then reinforce these social relations (Hayles 1999, xiii). This reverse feedback loop can be understood in terms of particular dramaturgies, as I will discuss in the next section of this chapter.

The drive for artificial intelligence and autonomous machines reflects and (re)inforces an idea of the human in which autonomy is privileged over relatedness (Suchman 2007, 213). Machine autonomy as idea and in technological form also seems to emerge alongside a shift in the location of political responsibility. On a mundane level this might be manifested in the ambiguities generated by the virtualisation through social media of aspects of everyday life. Within electoral politics, algorithms acquire political power through voter targeting and new ways of intervening in national politics across borders. More lethally, the development of autonomous weapon systems seems to be delegating the power to kill to machines and, in combination with a 'necropolitical' power to discriminate, reinforcing unequal global power dynamics (Mbembé and Meintjes 2003; Suchman 2015). Robots, in the form of unmanned autonomous vehicles (UAVs), i.e. drones, both signify and enact global power relations in which machines recognise which lives should be saved and which destroyed.

My argument in this chapter is that the theatre offers a site for exploring and intervening in such iterative processes of co-production. This is partly to do with the long history of theatre as a place for exploring, playing with and challenging modes of mimetic representation. However, because there is a significant overlap between mimesis in theatre and mimesis in computer science, I am not suggesting here that the theatre's political potential simply lies in representations of computers or computer scientists. As I argue in the next section, there has historically been a series of exchanges backwards and forwards between dramaturgies in computer science and dramaturgies in the theatre and popular performance which rather complicate the relationship between theatre and science. Beyond this though, I want to suggest that theatre holds potential for exploring the apparently proliferating and accelerating modes of existence co-produced with new computer and information technologies.

Theatre has an enduring capacity for what Charis Thompson has called 'ontological choreography', that is to say its various practices and institutions have been able to take what 'might appear to be an undifferentiated hybrid mess' and through particular choreographic approaches create 'a deftly balanced coming together of things that are generally considered parts of different ontological orders (part of nature, part of the self, part of society)' (2005, 8). In *My Square Lady*, the choreography deftly balanced a robot, a robotics lab, scientists, scientific research, opera singers, characters from operas, operas, musicians, music, production crews, Gob Squad, video recordings, algorithms, an opera house, a live performance and various publics with all their respective modes of existence.

The opening sequence of the performance of *My Square Lady* illustrated this choreography well. A technician walked onto the stage holding a microphone 'eins, zwei, drei, test'. He introduced himself and his role and, possibly more in hope than expectation, claimed that robots would not be able to replace him. The wardrobe mistress who appeared later was not so confident given that her sewing machine is already computerised. A procession of backstage staff deconstructed the technology of opera as labour under threat from mechanisation. The video screen interrupted with a recording in which we heard the roboticists tightening screws and checking batteries. Myon presented little threat as scientific equipment, easily disempowered by removing the battery. Music swelled, a choir of children dressed in homemade, cardboard robot costumes with technique and power rather cutting against their age sang 'sanctus, sanctus Dominus Deus…'. The simultaneously spiritually transcendent and cute parody was again interrupted by Myon's point of view on the screen (a recording but as if live) going up and down stairs backstage. Now slightly uncanny, the scientific instrument has become TV star backstage. Robotics researchers and opera singers entered; an adult choir bolstered the children's voices. Volume and spectacle were growing, brought back down to earth with Sabine the stage manager reminding us who really cued the interruptions. The climax… Myon appeared as camp *deus ex machina* through the floor and a cloud of stage smoke. The scientist, Professor Hild, stormed on and shouted 'Stop!' Theatrical spectacle suddenly gave way again to theatrical (clearly cued if only loosely scripted) engagement with science. Myon was a threat to workers' rights, a potential carer for an elderly relative, a God trick, a theatre prop, a fragile research instrument and more, held together by the choreography.

Of course this sequence was funny. Unlike the assisted reproductive technology clinics that Charis Thompson was studying, theatre does not have to take its ontological choreography seriously. Theatre tends to make fun of and with what it is reflecting. It provokes funny feelings and assembles publics for and through laughter. In fact its lack of seriousness, the laughter or at least the possibility of it breaking out, is central to the theatre's capacity to hold different kinds of things together.[4] Comic theatricality therefore offers ways of holding together proliferating ways of being through their very unknowability. Of course, if I am right about this, I am clearly setting myself up to fail. However, in this chapter, I also try to understand how this might work. Whether or not this is successful though, and failure may be as important as success, I also want to highlight how this choreography might constitute important political work even if it is not productive of knowledge in any accepted sense.

Robots in the theatre: biomimetic dramaturgy

Myon was not the first robot to go to the theatre. According to many histories the term robot was a theatrical rather than scientific invention, coined by Karel Čapek in his 1920 play *R.U.R.* (*Rossum's Universal Robots*)

(2015), within a longer history of automata and performing objects that has been clearly articulated in book form by theatre historian Kara Reilly and as a major touring exhibition curated by Ben Russell in 2017 (Reilly 2011; Russell 2017). In the dystopian narrative of *R.U.R.*, the robot characters that ultimately overwhelm their human developers have generally been played by human actors, at least until Leonel Moura's version in São Paulo in 2010 in which robot actors played Babá, Primus and Helena. The apparent historical significance of *My Square Lady* though was in the degree of autonomy afforded by Myon's particular technology, that is to say, its ability to make certain decisions according to its programmed algorithms but unpredictable to its human programmers. Myon could improvise.

Much of the excitement and critical debate generated by *My Square Lady* and by earlier comparable experiments emerged from the apparent ability or inability of nonhuman robots and automata to pass for humans or indeed for human robots to pass for nonhumans. In the eighteenth and early nineteenth centuries, the Hungarian inventor Wolfgang von Kempelen's chess-playing Turk appeared to be an automaton that could beat human chess players but actually contained a human operator. In a catalogue essay for the 2017 *Robots* exhibition, curator and historian Andrew Nahum explained the way this illusion worked to dramatise machine intelligence before its technical realisation was actually possible:

> It toured widely round the world but the device was an illusion. Each chess match was preceded by a presentation of the automaton and its 'machine room' in a performance that resembled the techniques of stage magicians. As the presenter opened doors, rotated the automaton and opened another door to show a fresh view, a concealed chess player manoeuvred around keeping out of sight. And before play commenced, the introduction ended with the showman theatrically winding up some noisy clockwork. (Nahum 2017, 60)

In the essay, Nahum draws attention to the way the performance provoked speculation and thereby drove research in particular directions, for instance by prompting early computer scientists Charles Babbage and Ada Lovelace, in the first half of the nineteenth century, to think further about the potential of machines to do mathematical calculations.

This biomimetic dramaturgy is enduring and powerful. The imitation of a human by a machine or a machine by a human generates tension and suspense in the theatrical event where the audience speculates about the specific technology behind the doors. Many multimedia artworks presented at international events such as Linz's Ars Electronica Festival rely on this dramaturgy. Such feelings permeate beyond the event though, raising excitement about its potential and fears about the extent to which machines may displace humans, not just at games of chess but in weaving, calculating and a variety of other laborious and wage-earning practices. The drama is not particularly reliant on a specific technology (hence its durability), but rather relies on setting up an ontological uncertainty. The question about the box's mode of existence is either/or: either it is a human or a machine. Paradoxically, the technology, whether or not there is a human body inside the box, both matters and does not matter. It matters in terms of the performance's intensity: the extent to which it engages. Yet, it does not matter in terms of its affects and effects: both the human as robot and robot as human effect speculation, feeling and power. As such the operation is somewhat different from the ontological choreography that emerged within the *My Square Lady* project. However, the project, in much of its publicity, articulated itself in terms of the established biomimetic dramaturgy I have just described.

Myon was introduced to the Gob Squad group and a range of singers and production staff at the Komische Oper in 2013. Ostensibly this would be the start of a two-year process to teach Myon to become a performer and, in many ways, would be a continuation of a biomimetic dramaturgy embedded in Myon's design and heritage. Many of the robot's features were designed based on a human analogue: its body shape with two legs, two arms and upright stance; its SpinalCord communication mechanism; a processing unit in the head called the BrainModule; and joints that worked through antagonistic actuation. It was designed specifically to model human cognition related to language use. This research focusses on how humans *evolve* the ability to understand and use language and is concerned primarily with the perceptual apparatus needed prior to sophisticated, linguistic operations (Steels and Hild 2012).

Language-based biomimetic dramaturgy within scientific discourse is closely associated with the extremely influential script provided by Alan Turing's 1950 paper in which he asked if a machine could think or, in his proxy formulation: 'Are there imaginable digital computers which would do well in the imitation game?' He proposed a particular way to stage this question in what has become known using Turing's term as 'the imitation game' or more commonly using his name, as the Turing Test (Turing 1950, 442). In this game or test, computer scientists are challenged to design a programme that, in some form of conversational interaction, can deceive more than 30 per cent of a group of people about whether it is a machine or another human. Such stagings of Turing's script generally figure the human as a particular kind of plausible, logical rationality. The test is usually administered by text on a screen through conversational questions and answers posed by software known as a chatbot or chatterbot.

For instance, the Eugene Goostman programme managed to fool 10 out of 30 judges in a test in 2014 by pretending to be a Ukrainian teenager with a pet guinea pig (Sample and Hern 2014).[5]

Myon's biomimetic design, and that of other comparable experimental humanoid robots, attempts to restore other kinds of sense perception as the basis for and medium of communication. Myon is not reliant on a text-based interface and can sense a range of visual, audio and kinaesthetic stimuli. The robot therefore seems to restore the body lost within Turing's dramaturgy. However, to accept this proposition at its 'interface value', to use Sherry Turkle's (2012, 18) elegant turn of phrase, is to miss the way that new media dramaturgies recycle old media dramaturgies such as the fake chess-playing Turk. Even as the biomimetic potential of such technologies extends beyond text on a screen, it appears that the dramaturgies, and therefore the co-produced social relations, are persistent. What might a creative research process look like that is both reflexive towards the tendencies of artificial intelligence to revert to the imitation game and opens up the possibilities for reworking the dramaturgy? It seems to start with a sense of humour – a laugh at the technology and our interactions with it – and a theatrical character, Eliza.

The pioneering computer scientist Joseph Weizenbaum bemoaned the tendency of the public to take his work too seriously rather than not seriously enough and turned to Eliza for help. Weizenbaum worked on artificial intelligence programmes from the early 1960s. A number of scientists had devised programmes that could respond plausibly to questions posed by humans via a text-based computer interface, but, as Weizenbaum observed, 'understanding must be mutual in most realistic situations'. He therefore set out to create a programme that could ask questions rather than just answer them. The script he chose was based on that of 'a non-directive psychotherapist'. He chose this scenario as it worked towards a simulated conversation, what researchers called 'natural language', but avoided the demand for the machine to have extensive 'real-world knowledge'. He figured that, as a therapist, the software could reasonably just bounce questions back at its interlocutor. Drawing similar comic inspiration to Gob Squad, albeit 50 years earlier, he called the programme Eliza, after Eliza Doolittle, explicitly signalling his work's theatricality referring to G. B. Shaw's play *Pygmalion* as his source. Although it was ground-breaking at the time, Weizenbaum was very aware of the programme's limitations and indeed sceptical about the machine intelligence project more broadly. He thought this was clearly demonstrated by Eliza's performance. For him, Eliza's conversations were a 'parody' of a real conversation. However, he was repeatedly frustrated. People, he observed, seemed to take the programme seriously. They would 'insist, in spite of my explanations, that the machine really understood them' (Weizenbaum 1976, 188–9).

As his career developed, Weizenbaum, who might have been expected to be an enthusiast for a technology he played an important role in developing, had increasing political reservations about the direction the science and technology was taking. This seems to mirror the trajectories of scientists involved in the Manhattan project, notably Joseph Rotblat who became a leading anti-nuclear activist, geneticists such as John Sulston who became increasingly concerned with biomedical ethics, and other scientists like Stephen Hawking who expressed growing concerns about artificial intelligence. Weizenbaum's trajectory is documented in a remarkable film by Jens Schanze (2010) in which the elderly scientist's concern is juxtaposed with the futurist jingoism of other scientists such as Raymond Kurzweil and Hiroshi Ishiguro. At the end of the film, which features the end of Weizenbaum's life, he comes to terms with what he regards as a pathological tendency to attach to computers through his experience of mortality derived from listening to Bach in his apartment.[6] As therapeutic as such transcendent spirituality may have been late in his life, I think he was also onto something with his earlier concerns about public tendencies to take the technology too seriously. I suggest his attempt to frame his experiment as comically theatrical was the right impulse but misfired, at least in part, due to problems with its staging, which I think Gob Squad's project went some way to addressing.

Part of Weizenbaum's difficulty was that the sciences of robotics and artificial intelligence, as well as a wide range of sciences that draw on cognate practices, are already mimetic. They do not therefore function according to the rules of 'normal science' as articulated by Thomas Kuhn (Kuhn and Hacking 2012). For Stengers, such new modes of research are reliant on simulation. Researchers in these fields are not referring to 'a truth that would always silence fictions'. Rather, they are using modelling and forms of 'mathematical fiction' to reproduce phenomena, whether that be human thought, motion or indeed the global climate (Stengers 2000b, 135). Within Kuhnian normal science, the paradigm gives coherence to scientific programmes. It stabilises norms, institutions and disciplines that produce scientific truth. Stengers argues that this 'form of mobilization' and indeed the paradigm itself may be giving way to 'the speed of technical innovation' itself (Stengers 2000b, 116). Before the paradigm has stabilised, the amount of data generated and the technological means by which the data is gathered have made it obsolete. Mimetic fictions in themselves, for instance in Weizenbaum's staging of his technology as Eliza, do not necessarily question or complicate such scientific practices. The political challenge here lies not in how to uncover the reality of things by exposing an underlying layer of (social or physical) materiality. Rather, it has to do with how publics form and fashion relationships with and in relation to such objects.

The resources of theatricality offer the possibility for orientations towards objects that are in some respects unreal or not yet real. As Rebecca Schneider has pointed out:

> Theatricality, unlike performativity, makes sure we remember that not everything in the world is real – or not *only* real. The matter of mimesis, in both becoming and unbecoming, might rather (or also) be posited as agential theatricality: becoming unreal. And as many have said, theatre – the playground of whores and dandies – is *really unbecoming*. (Schneider 2015, 14)

Weizenbaum's problem though suggests that theatrical mimesis alone may not necessarily be all that unbecoming: the tendency in the biomimetic computer sciences is for technologies (too) rapidly to become real. As the theatrical is constantly subsumed within social and scientific processes it does not always generate the kind of laughter of truth that Stengers has called for. If dominant conceptions of public engagement with science attempt to *real*-ise science, i.e. enable transfer into technology, public acceptance or understanding, I suggest that comic theatrical engagement in the theatre can be more a kind of unrealisation. This is not the same as a Brechtian estrangement device that reveals an underlying (dialectical) materialist reality; neither is it about an appeal to an essentialised idea(l) of the human under threat by modern science or technology. Rather, what comic theatre may have to offer is processes through which premature realisations or naturalisations can be suspended, where simplistic oppositions between truth and fiction, objectivity and subjectivity can be scrutinised and indeed where new technologies may be reconfigured.

When Weizenbaum named his programme Eliza after Eliza Doolittle, he was acknowledging the theatricality of his scientific experiment. However, his idea of what constituted this theatricality and indeed his reading of Shaw's *Pygmalion* are worth interrogating more closely for the extent to which they might have led to a public response that he regarded as too serious. The parallel he drew between the software and Miss Doolittle was based on the idea that they both 'could be taught to "speak" increasingly well' and that this could be done by training them to follow scripts (Weizenbaum 1976, 188). However, where the analogy fell down, significantly, was in his figuring of Eliza as 'an actress… who had nothing of her own to say', in his conception of theatricality as 'suspended disbelief' and indeed in his omission of key aspects of *Pygmalion*'s dramaturgy (Weizenbaum 1976, 189).

Weizenbaum's staging of his software effectively reduced the narrative to the scene – in *Pygmalion* it is an ambassador's reception – where Eliza attempts to deceive the assembled aristocratic socialites as to her origins. In most stagings of this scene, Eliza's success is demonstrated by the admiration she attracts from the party guests and the conclusion reached by another linguistic expert at the party that she is a Hungarian princess. None of the characters she meets at the party realise that she is actually a Cockney flower-girl. In restaging this scene with his software as Eliza, it should therefore have come as no surprise to Weizenbaum that it was similarly successful. What Weizenbaum did as dramaturg though was to narrow the theatrical frame focussing attention onto Eliza and the assembled public who may or may not be deceived, restaging Turing's imitation game or indeed von Kempelen's Turk. In Shaw's original, this scene takes place in the middle of the play, between Acts 3 and 4, if it is staged at all.[7] The play and indeed most of its adaptations concern themselves mainly with the design of the experiment, Eliza's interactions with Higgins and the other characters, and the unravelling of the implications of Eliza's performance at the ambassador's reception for both her and the other characters. Of course, in this way, the public, as in the theatre audiences, are not positioned as potential dupes of Eliza in an imitation game. They know she is not a duchess and are repeatedly reminded. The drama of the play rests not on the performance of Eliza as a duchess, but rather on her developing capacity to pretend and the interactions she has with Higgins and the other characters. Comic theatricality often emerges in performances of *Pygmalion* from the audience's awareness that it is an actor playing Eliza playing a duchess.[8]

Despite the tongue-in-cheek project publicity and the tendencies of artificial intelligence to revert to its core dramaturgy, *My Square Lady* was similarly careful not to restage the imitation game or create the illusion of Myon as human performer. During the project, the robotics team were not just representing or showing off the results from an experiment but actually doing one.[9] In workshops and rehearsals, this was constantly evident as scientists tinkered with the robot and otherwise intervened in the process. Performance audiences were repeatedly reminded that they were watching a scientific process. Bastian Trost from Gob Squad introduced the performance piece as 'a real experiment' rather than a game or trick; there were conversational reflections on the implications of Myon's technology for various other participants and projections of video clips from the two-year process.

Of course, in *Pygmalion* there are two main characters Eliza *and* Higgins who both undergo a kind of transformation (or what Shaw called 'transfiguration') in the course of the drama. The ur-text in Ovid's

Metamorphoses is as much a story about the sculptor Pygmalion's creative process as it is about the statue that is brought to life by the goddess Venus. Artificial intelligence, identifying itself as a computer *science* through its apparent objectivity, is inclined to present its results. It rarely shows all its representational apparatus, that is to say the scientists and various technologies that go towards staging its imitation games, or the effects of these results on that apparatus. In his presentation of the Eliza software, therefore, Weizenbaum removed himself (Pygmalion/Higgins) from the theatrical frame.[10] *My Square Lady* reinstated Higgins by incorporating Professor Manfred Hild and a number of his research team not just into the creative process but also on stage in the final performances, as I will explain in more detail later on. The final performances were really an episodic documentation of this process which itself reflects Eliza's education in *Pygmalion* or *My Fair Lady* and the subsequent exploration of its repercussions for all those involved.[11]

Importantly though, *My Square Lady* went somewhat further than either *Pygmalion* or *My Fair Lady* in that Myon, as Eliza, whenever invited to play the imitation game, tended to fail to perform or, at the very least, fail to imitate or fail to impress. In the performance, a screen above the performers' heads displayed sometimes a 'live' feed from the camera in Myon's head, so Myon's 'point-of-view', and sometimes a recording made using Myon's camera but played back on cue via a different video technology, presumably operated by a stage technician, although this is not made clear, so is deceptive in some sense. For example, in one sequence of the final performance, a recording gave a clip from a series of rehearsals involving Arno Waschk the conductor trying to teach Myon to conduct. An image on the screen also showed Myon's 'memory' of being taught to conduct. The algorithm that detects and reproduces the reoccurring movement motif was represented as a stick figure conducting. Myon was then expected to perform, bringing in the orchestra for the 'Brindisi' chorus from *La Traviata*. This was staged with a great air of anticipation and build-up. Myon at first failed to begin at all. Waschk and Sean Patten from Gob Squad chatting on stage speculated on why. In a series of ironic anthropomorphisms they wonder is it 'Lampenfieber' ('stagefright'), or is Myon thinking about it too much rather than getting on with it? Waschk helped, guiding the robot's hand. Then with a cue from Torsten Siedel, researcher and Myon's bodyguard and handler, Myon managed a halting performance and stopped. The cast responded with neither enthusiasm nor disappointment. The audience apparently registered both: they laugh *and* applaud. Patten, deadpan: 'Das müssen wir erstmal sacken lassen!' ('Now let's let that sink in!') (Figure 2.1).[12]

Figure 2.1 Myon conducting in *My Square Lady*

The staged and felt anti-climax here and elsewhere in the process as well as in the final performance was central to both the politics and aesthetics of the practice. I will address the detail of this feeling later; however, the building up and letting down through Myon's failure to perform was an important feature of *My Square Lady*'s dramaturgy. Within the experimental context it was not (just) a failure in the imitation game but a failure to play it. It engaged with the imitation game as script, but by not simply enacting it, it presented a challenge to or disruption of the prospect of an ontological shift that Auslander (2002) argues is presented by digital technologies in general and the chatterbot in particular. Staging Myon alongside Gob Squad artists, opera singers, video feed, scientists and choirs constituted the ontological choreography in which different ways of performing were juxtaposed and held open to scrutiny. Steve Dixon (2007, 273–4) has helpfully theorised failures of robot performance as 'metallic camp' recognising their funniness and queering of the object. However, this seems to rely on figuring the robot as 'performative[ly] self-conscious' – the robot is granted an individualised, proto-subjective agency to 'camp about' at least within a particular set of theatrical conventions. Myon even fails to camp about. In one of the few pieces of dialogue it utters, it is 'sitzen[,] passiv' (which could be 'sitting passive' or 'sitting, passive' as Myon does not punctuate). The humour in *My Square Lady* arose from the interactions between a variety of different objects and their choreographed intrusions into each other's ontological orders.

Edward Scheer (2015, 143) has argued that robotics constitutes a kind of 'new media dramaturgy' and that the theatre might offer a space for debating human engagement with objects, trying out empathic relationships and forms of kinship. His thought-provoking analysis of the performance installations of media artist Mari Velonaki draws on the new materialist philosophy of Jane Bennett in articulating the vitality of the robots in Velonaki's work. According to Scheer, Velonaki's robotic dramaturgies, such as the *Fish-Bird* series, expose robotics as an 'agentic assemblage' in Bennett's (2010) terms. Here, agency emerges from the interaction of subjects and objects in relation. Velonaki's particular dramaturgy utilises a form of biomimesis that seems to mirror animal behaviours but stops short of the kind of visual mimesis used in humanoid robotics that might include Myon and more directly critiques the robotics of Hiroshi Ishiguro with its emphasis on facial and bodily appearance as a route to passing as human.

In many ways *Pygmalion, My Fair Lady* and *My Square Lady* seem to follow this approach to new media dramaturgy. Action and agency in *Pygmalion* emerge from an assemblage of human and nonhuman actors or actants including Eliza, Higgins, Pickering, Mrs Higgins, the phonograph, the embassy ball, Higgins' slippers, etc. The argument between Eliza and Higgins after they return from the ball as to who (or what) should take credit for her success in the imitation game emphasises the complex politics underlying the constitution of the assemblage. As Eliza hurls them across the room at him, Higgins' slippers are enrolled in her ambivalent transformation into Edwardian New Woman. At the start of the play Higgins' relationship with his phonograph constitutes his power 'to turn her on as often as you like' (Shaw 1987b, 36). By the end Eliza is able to turn the tables as Higgins confesses that he might miss her. She tells him, 'when you feel lonely without me, you can turn the machine on. It's got no feelings to hurt' (Shaw 1987b, 127).[13]

In *My Square Lady*, the assemblage is reconfigured in a number of ways, perhaps most literally by Myon's disassembly and reassembly frequently in rehearsals and notably at a key moment of the final performance. As mentioned above, Myon is a modular robot that can be taken apart and reassembled during run-time. This feature is apparently determined by a number of material constraints in the laboratory. As researchers are unlikely to need the whole robot for much of their work, the possibility to share the robot among a laboratory team, by dismantling it into arms, legs, head and body, makes for economic efficiency. For the audience of the performance and the theatre-makers in rehearsals, Myon is literally dismantled and reassembled: not one entity but an assemblage of pieces brought into being by a laboratory made up a number of researchers and limited financial resources. In performance, this process is staged as a tableau of Leonardo da Vinci's *The Last Supper* with Myon enrolled as Christ but then taken apart and passed down the table in pieces.[14] His disassembly and reassembly becomes an analogy for divine transformation directly echoing the idea of transfiguration used by Shaw. Rather than copying life, Myon is figured as god-like, immortal. And yet this is clearly tongue-in-cheek, Myon's head camera remains on throughout. When I watched this scene in rehearsal the tension on the faces of the researchers was clearly not so much about whether Myon would still work, perform or live as a whole entity after being reassembled. They do this all the time. It was rather to do with having to stand the rather unstable assemblage on the table. Any action – standing up, worrying scientists – here clearly emerges from the interaction between joints, sensors, algorithms, gravity, table, roboticists, etc.

This account is politically productive in constituting a feminist critique of scientific practice, underlining the importance of human relations and cultural scripts in the construction of technologies such as Myon. Myon's much vaunted autonomy is less remarkable than the operation of the assemblage. Myon does not act alone: it is neither the work of one scientist nor even of a whole team of scientists, but rather of a complex network of relationships. Such analysis also registers an ecological concern with reimagining and reconfiguring relationships with the nonhuman, whether this is Myon's plastic and metal components or indeed with the other species that reside inside and outside the human body. Lucy Suchman, Claudia Castañeda and Donna Haraway have all written extensively about the similarities between the way scientific research makes mimetic use of nonhuman species and machinic 'model organisms' (Castañeda and Suchman 2013; Haraway 1991). However, a new media dramaturgy that looks to foreground relationships and background the entities themselves seems to miss political and aesthetic features of *My Square Lady* that I found the most striking and funny. Myon is both unsurprising in its liveliness and, considering the scientific and theatrical build up, surprisingly unlively. Clearly it is the relationships that are the focus of the theatre, and yet the relationships are evidently co-produced with this specific object.

For objects to appear lively in the theatre or even within broader theatrical cultures, of course, should not come as a surprise. Fascinating historical accounts of this liveliness can be found in the work of theatre historians such as Kara Reilly's work on automata and Andrew Sofer's discussion of props (Reilly 2011; Sofer 2003). What is surprising and funny about Myon was that it often failed to come to life or at least stopped short of doing so. It was repeatedly invited to spring to life through the anthropomorphic gestures of the human actors, but, repeatedly, the choreography suspended the transfiguration through some kind of reconfiguration. Of course this resistance can be understood as an act of the (social) assemblage too. However, comic theatricality in *My Square Lady* proceeded not just from Myon's capacity to be lively or life-like, nor just from human tendencies to accord liveliness or subjectivity to nonhuman things, nor even for related tendencies of humans to objectify themselves and each other. Somehow, *My Square Lady*'s choreography tended to hold these tendencies up both in the sense that it presented them theatrically and in the sense that it stopped them momentarily, resisting the possibility for simple conclusions to be drawn about the limits of subjectivity or objectivity. This choreographed hold-up is not the same as the tension within biomimetic dramaturgy where engagement emerges from ontological uncertainty. Rather, such theatrical choreography configures and reconfigures different relationships between humans and nonhumans and explores their different modes of existence.

In the re-enactment of the Last Supper described above, Myon, as Christ, seemed about to be transfigured and is accorded posthuman subjectivity. This process is made sense of through religious imagery as a reconfiguration of the divine. At the same time, though, Myon was taken apart, with its continuing capacity for visual perception clearly mechanical: Myon made sense as a video camera on shaky legs. The humour here is literally but comically iconoclastic in that it emerges from the making and then breaking of likenesses: Myon as Christ, Myon as video camera. Laughter at this moment registered the way that both configurations made sense, that Myon could not be reduced to either and that this situation was complex, troubling and defied more coherent articulation. Being with Myon, as is the case with many technologies, was thus both engaging and exhausting. It was this exhaustion, as much as any mimicry, that provoked a lot of the laughter in rehearsal and performance.[15]

In rehearsals a lot of laughter and bemusement among Gob Squad members, opera singers and scientists emerged from exercises in which Myon and various performers interacted producing unexpected or surprising results. Often these results could neither be easily explained in technical terms nor through a particular socio-cultural reading. There were a number of sessions in which Myon reproduced singing exercises modelled by opera singers. These often came out in a way that was both identifiable with the performance of the singer and also apparently distorted or otherwise acoustically transformed. Accounts of this were offered that made sense of it through science fiction tropes – a Gob Squad member likened one of the sounds to the distinctive, unearthly soundscapes of the well-known British television drama *Dr Who*. Singers made sense of the phenomenon through their understanding of their own singing technique: Myon was attempting to reproduce vibrato. At the same time the scientific team attempted to explain the distortion through the various ways that Myon might have been reproducing the audio input: for instance through the software's propensity to re-synthesise the sound according to a different algorithmic process than other kinds of analogue or digital audio recording equipment. Collectively, though, they made sense of the moment and their perplexity through laughter.

In performance, the soprano Mirka Wagner took Myon on her knee like a child and, with immense power, sang an aria from *Rusalka*. She gave up. Myon was too heavy, scratchy and unwieldy to be a child and was unmoved, at least not until a couple of the researchers came to her rescue to get him onto and off her lap. In another scene which supposedly demonstrated Myon's human-like capacity to have its attention attracted by a variety of (particularly red) stimuli, Myon registered barely a flicker on the point of view screen as world-class opera singers – professional attention-grabbers and collective products of years of training, technique and experience on this and other stages – brandished a variety of red props in a three-pronged act of seduction from *The Magic Flute*. As already described, despite extensive training from the musical director, a metronome would conduct better than Myon.

Myon's scientific properties were fascinating and theatrically generative despite and perhaps because of an apparent lack of action which limits the potential of an analysis of the practice through notions of biomimetic dramaturgy. For object-oriented philosopher Graham Harman, 'to treat objects solely as actors forgets that a thing acts because it exists rather than existing because it acts' (2016, 7). Technical accounts of objects tend, in Harman's terms, to 'undermine' by reducing them to their components. Socio-cultural accounts may reveal the way objects are socially constructed according to gender norms, in a way that is determined by economics, or through religious or secular accounts of human exceptionalism. For Harman, such critiques have a tendency to 'overmine' objects, reducing them to a particular account of the social. It is not that these approaches are not useful but that they do not exhaust the possibilities of an object. There is a 'surplus of reality' (2016, 10) that exceeds what objects actually do. It is this 'immaterial' surplus that accounts for Myon's 'surprise and opacity' and which the project's ontological choreography explored (2016, 19). The project was notable in the way it balanced and juxtaposed tendencies towards overmining and undermining Myon. Such leaping between different and potentially incompatible accounts was labour intensive, involving the resources of opera house, laboratory, art collective and various audiences. It was also exhausting and, in many ways, undramatic. The juxtaposition of contradictory explanations, the labour-intensive process and indeed the sheer exhaustion resulted in laughter, and laughing turned out to be a serious response to Myon's surprise and opacity.

Emotion capture: doubling sensation

In an interview I had with Gob Squad member Sean Patten towards the end of the rehearsal period, he told me that part of the driving force behind the project had been the fascination many of the collective found for popular science fiction cultures. I then asked him about his feelings after two years of working with a real robot and expert robotics scientists:

> I feel I can sleep safely at night knowing that Terminator isn't quite round the corner and this is the funny thing about research into AI and things like that, that it is always in 30 years' time. You know they are going to take over the world and everything like that. It just shows that scientific development always happens in a different way than you expect. We always look at the robot and think 'Is that Terminator? Is that C3PO?' But a lot of those things you might call artificial intelligence have crept up on us in a very different way. You know through cookies, algorithms, smartphones that can track where we go. That's the sort of networked AI rather than something that inhabits [a] two arms and two legs robot.

It's one thing to understand or have dramaturgical mechanisms for conceiving of or engaging with (novel) objects or ontological choreographies that juxtapose a robot as social object and proto-subject. It is another to conceive of a political position towards this object, a way of recognising a position of responsibility in relation to such an object. Patten's reflection registered the ambiguities and contradictions of this situation and seemed to echo the contradictory position experienced by Weizenbaum I discussed at the start of the chapter.

His first response here suggests a sense of relief that the dystopian imaginary of the Terminator is not likely to come to pass in the imminent future. The anti-climactic aesthetic of the performance piece and Myon's failure to perform would seem to reflect this disillusionment. Patten passed on this framing of Myon as disappointing, letting the audience down gently near the start of the performance:

> Sean (to audience): Myon is a REAL robot, not a puppet, so those of you expecting lasers and mechanical acrobats, might be a little bit disappointed. There's the door.

This kind of reveal – showing the reality behind an apparent illusion – seems akin to Weizenbaum's fruitless attempts to explain and therefore articulate the limitations of Eliza. Patten distances Myon and the real robotics from the dystopian robotic imaginary conjured up by films like *Terminator*. It is a persistent trope within science communication to respond to popular hype – the articulation of heightened feeling in the media often related to robotics or genetics – with deflationary attempts to tell the real story. This is often doomed, as Weizenbaum found. This mode of science communication was directly referenced in performance as Manfred Hild, after his 'Stop!' interruption, launched into the familiar TED talk pedagogical aesthetic explaining the reality of his research as a response to the appearance of Myon through the floor as theatrical *deus ex machina*.

According to Alenka Zupančič, this kind of reveal is an ideological mechanism of power and control:

> There is something in the modern ideological structuring of knowledge (which often takes the form of revealing what is 'behind the scenes' – before anybody even thinks to ask) that seems to absorb the question of consequences and to make it literally meaningless, incomprehensible. (Zupančič 2016, 413)

Revealing the real in this way then, rather than fostering insight, tends to obfuscate. The reveal by Hild and Patten though is theatrical. It is staged not to offer privileged access to real robotics – it is not that kind of show – but as a comic double take: a parody of the realist reveal as aesthetic or political strategy and a way of registering both their perplexity and the way that Myon exhausts their different ways of making sense of it.

Patten's interjection quoted above was itself one of a number of interruptions he and other Gob Squad performances make to Hild's TED talk. 'This isn't lange Nacht der Wissenschaft, this is the real thing' remarked Patten in one of them. In this case the 'real thing' was not the 'long night of science' – a popular science communication event format where scientists spend an evening explaining their research to the public. The real thing was 'showtime'. So this was not the procedural aesthetic of the science documentary, where scientists explained how things really work, any more than it was an attempt to deceive the audience by theatrical illusion. Hild interrupted any attempt to overmine Myon by presenting it, however theatrically, as divine. Patten interrupted Hild's equally theatrical attempts to undermine Myon through science communication. We were caught in an epistemological stalemate and it felt funny.

In media production, the technology for turning human bodily motion into digital data is known as motion capture. Typically, in film, TV or video game production, sensors are attached to actors' bodies which record movement in space or special cameras track markers overlaid on a human or animal face. This data then enables the reconstruction of 3D computer models of the movements that can then be (re)animated in films or games. Myon makes use of similar technologies with an in-built set of sensors that enable the capture of motion through touch, vision, audition and then its reproduction. A conceit at the heart of *My Square Lady* was an experiment in emotion capture: the question as to whether Myon could be moved by opera. Myon captures (e)motion in at least three forms: as represented human emotions (nostalgia, desire, anger) through recording images and sounds of operatic scenes; as behaviours and physical skills such as singing through the audio recording and reproduction and conducting both through visual and haptic stimuli when Arno Waschk actually held Myon's hands; and also sets of public feelings about Myon as technology and the sciences of robotics and artificial intelligence more broadly as visual image, audio recording and to a lesser extent manipulation and touch as when Mirka Wagner tried to hold the robot in her lap.

Brian Rotman argues that motion capture technology marks a particular historical shift – an ontological break that has also been characterised in terms of the posthuman – comparable to the introduction of alphabetic writing. 'Gesturo-haptic writing', as he calls it, enables the encoding of the non or pre-verbal (2008, 54). According to him, alphabetic writing has had a series of profound theological, physiological and ontological effects including monotheisms and the mind. The gesturo-haptic, he implies, holds similar potential. It ushers in an experience of what he calls a 'para-human' condition – a feeling of being beside oneself. Rotman cites Artaud's Theatre of Cruelty as the theatrical pre-figuration of the gesturo-haptic: a turn to an immersive theatre of raw sensation, a theatre potentially without written or even verbal language.

In *My Square Lady*, sensation was often doubled rather than being raw or experienced through immersion. People were moved and then the movement was captured within another medium or theatrical ontology. Things appeared beside themselves. There were repeated rapid juxtapositions of sensation with reflection of and on this sensation. The doubling of sensation inheres in the ambivalences and ambiguities of science in performance. *My Square Lady* captured the sense of movement between states of being rather than proposing a novel sensation or unified performance ontology. The live comments and recordings of the opera staff's responses to Myon quoted at the start of the chapter were characteristically ambivalent about the hopes and fears engendered by the technoscience they saw Myon as standing in for. Myon's theatrical appearance as both or alternately amazing wonder and technological failure articulated this ambivalence. The stage technician hoped that robots would make his work easier in the future but feared thereby being made redundant. The violinist was convinced that she could not be replaced by a robot even as the piece itself featured music generated by a range of acoustic, analogue, electronic and digital devices.

In the performance, as the cast assembled into *The Last Supper* tableau, casually passing a loaf of bread between them, breaking off pieces and chewing, shortly before Myon was disassembled, Sean Patten, Bastian Trost and Johanna Freiburg from Gob Squad fired questions about emotions at each other and various others onstage and off:

> Johanna: Sean, was hast du in deiner Familie über den Umgang mit Emotionen gelernt? [Sean, what have you learned from your family about coping with emotions?]
>
> ...
>
> Sean: Bastian, tell Myon all the things that make you personally cry.

The baritone Bernhard Hansky in a white suit was wheeled onstage in a sparsely furnished, panelled room. He started to sing Gluck's aria 'Ach Ich habe Sie verloren'. The context of this aria in Gluck's operatic version of the *Orpheus and Eurydice* story conveys the dangers of showing feelings although this was not explained in the performance: Orpheus kills his love by looking back at her. It expresses the sense of loss and also to some extent represses it. It is already an ambivalent expression of loss in a 'happy C-major mood' (Abbate 2001, 14). In Hansky's version, the loss was expressed initially as a controlled still pain. He barely moved and the power of his voice was held back. Opera scholar Carolyn Abbate has drawn attention to the 'automaton' quality of Orpheus in this particular operatic treatment of the story and I will return to the choice of a baritone to sing this aria that is more normally sung an octave higher by a mezzo-soprano. As the tempo slows, he grasped hold of a bunch of flowers from a vase on a small table. It was like a still life: a bottle, a violin, a vase with some flowers on a small table. His stylised rage grew: he threw the flowers at the wall, then the vase, then the bottle which smashed. He smashed the violin – there was gasp in the audience – picked up the chair and crashed it into the floor as he was wheeled into the wings. Patten meanwhile began to relate a story of when his bottled-up feelings exploded out in an unlikely confrontation with a doorman. Johanna Freiburg asked tenor Christoph Späth to simulate crying for Myon, which he earlier claimed to be able to do. Myon's attention on the point of view screen switched from the table, to Späth's face, to the floor.

Rather than gliding over the ontological break into an aesthetic where motion capture is smoothly integrated into a seamless, immersive experience, as is often the case in Hollywood cinema or even in more experimental work such as that by Velonaki, *My Square Lady* hopped backwards and forwards across the break, both generating and exposing a series of contradictory feelings, experienced within different theatrical modes of being. Hansky as Orpheus controlled a sense of loss, safely smashing a bottle, boxed into a stage on a stage, boxed into a score and a libretto, experienced as metaphor performed and received via 200 years of opera history. Hansky as Hansky then stepped out of the box onto the stage and confessed that he had never really experienced the feelings as he just attempted to reproduce them. Patten as Patten confessed to metaphorically bottling up anger and to an episode of its violent release. Späth attempted to illustrate for Myon the fakeness of the theatrical (e)motion by appearing to cry.

Metaphors were repeated and in their doubling lost their metaphorical quality as substitute. The bottle on stage might perhaps have stood in for repressed despair or rage. It was smashed by Orpheus (as theatrical, enacted metaphor), it was reassembled (as narrative metaphor) by Hansky, it was smashed again (as narrative metaphor) by Patten. The theatrical doubling of the metaphors started to turn them into metonyms, associations rather than substitutions: Orpheus' bottle and Patten's bottle, the stage on the stage, Hansky as Orpheus and Hansky as Hansky, Späth's tears on his face, Späth's tears on Myon's memory card. They all stood next to each other rather than stood for each other.

My Square Lady clearly registered a complex and probably representative range of public feelings about AI and robotics. In performance, these were reconfigured through repetition and metonym. This is a clear departure from dominant cultural modes of engendering affect through (robotics) science in performance such as the enduring *Frankenstein* tropes that express and drive a fear of runaway science and technology or attempts to articulate a more hopeful vision within the marketing cultures of technology multinationals and some documentaries. It also diverges as a practice from some utopian scenarios within experimental arts and science practices from Ars Electronica's attempts to rehabilitate the drone as harmless, celebratory spectacle within their Spaxels' performances to Ishiguro's Geminoid robots (and their appearances in the theatre). While *My Square Lady* prompted critical thought about robotics science and stimulated ambivalent feelings of hope and fear in response to the science and technology, it did not *simply* do either of these. The discontinuities, interruptions, anti-climaxes and repetitions already outlined all tended to produce comic effects in a process that tended to subvert big feelings of terror or joy in favour of smaller more nuanced affects.

In Zupančič's terms, comedy does not create intellectual distance from our feelings but creates a distance or lack of immediacy – hence the discontinuity – within these feelings. It is not about a critical distance from a fear of robotic technology but a critical gap within that fear. This gap can be understood spatially or temporally. From Captain W. H. Richards' robot Eric who toured the world in the 1920s to RoboThespian now available from Engineered Arts for public purchase, humanoid robots are frequently figures of fun as much as they are frightful figures of the posthuman. Part of the power of *My Square Lady*, as evidenced from its wide media coverage, was its novelty: the sense that a robot at the opera is peculiar, quirky, unexpected and yet somehow capturing the contemporary imagination. The juxtaposition of a cutting-edge shiny plastic robot with opera singers, whether or not they are in eighteenth-century dress, and even in a well-equipped technologically sophisticated opera house, still generates a curious sense of anachronism and yet it is not clear what or who is out of time.

As much as it was made clear in the project's publicity and within much of the dialogue between artists and scientists that Myon was a cutting-edge contemporary research robot, in many respects the feelings it generated emerged from its apparent anachronism. Myon appeared outdated within a contemporary cultural vocabulary dominated by a particular biomimetic dramaturgy. On stage, discounting the barely visible shifts of attention and wobbly balancing, Myon did not do anything more (and in many respects did less) than its apparent predecessors with their various combinations of human and machine: from Von Kempelen's mechanical Turk to C3PO familiar to *Star Wars* fans since 1977. Negative feelings about technosciences often respond to a sense of disorienting and uncontrollable speed, and yet here robots in science seemed to lag far behind robots in popular culture. This sense of Myon as outdated was highlighted in *My Square Lady* as Myon sat passively amid a stream of eclectic popular cultural reference points. In the opening scene referred to above, it was surrounded by the children and adult choirs dressed in historical-futuristic robot costumes that it clearly resembled. Myon was presented as a futuristic object and also as an object of the past, somehow temporally doubled both ahead and behind its time. This doubling created a certain jamming or slowing effect on the experience of technological speed that is a recurring theme in this book.

The temporal doubling was facilitated by the two different theatrical temporalities that co-existed within the practice. Opera time is governed by the linearity of the music – a tempo and duration that may be 200 or more years old. Action needs to occupy the time taken to play a particular section of the score. For this reason, also the staging needs to be fixed a significant period of time before the performance. For Gob Squad, spontaneity is a key dimension of their approach to making theatre, so changing things at the last minute is often an important aesthetic strategy. Sarah Thom commented to me:

> We had to do what's called a stage rehearsal, where we're looking at the stage elements that we're going to use and we did that four months ago. Usually with Gob Squad… where are we now? We're three weeks before the show. We'd be starting to pin things down. We've been pinning this down like a year ago… when you see it you go that's a bit clunky, and that's not flowing and these two there's a link there that we hadn't seen before. We can't do that because there's a fucking orchestra with notes written. You can't say 'do you know what, all those notes that took someone a month to write just chuck them away'. You just can't do that. So for us it's like whoa. Forward thinking.

For robotics, there is both a sense of rapidly changing technology but also the sense that the duration of the project may last decades. Myon's data and algorithms (just like Gluck's score) have taken years of development and are likely to outlast the humans doing the programming or composing. However, the exciting prospect of the project for Manfred Hild was not so much Myon performing scripted (i.e. programmed) actions on cue, but rather the potential for Myon to receive various stimuli and react unpredictably:

> What became very important was that we would not be interested so much in introducing the robot into a theatre play with a script. I mean normally a theatre play is a scripted thing. Everyone knows when everybody enters and leaves the stage and sings what and when and in which sequence. And, of course, this is most boring for a researcher and would have given us a hard time. And so we agreed upon that we would try to let the robot live within the piece on its own like stimulus-response… but not pre-programmed what happens next.

The most labour on the part of the scientists and Gob Squad went into creating the relatively unspectacular (illusion of) spontaneity most visible in the unpredictable shifting of Myon's attentional frames, that is to say where Myon's head camera is focused. Cued action would be both harder and less interesting. In both respects Myon underwhelmed and appeared out of time, accustomed as contemporary audiences are to the spectacular cued performances of robots.

Anachronism and ambivalent feelings of attraction and repulsion, according to Sianne Ngai, are characteristics of the gimmick as aesthetic category and emerge from the gimmick's relationship to capitalist production. The gimmick is 'both the effect and the procedure used to generate it' (2016, 485). It is both labour-saving and labour-replacing and therefore has the effect of shortening time. It is this abbreviation that generates the kind of irritation registered in a number of the responses to the robot quoted already when reflecting on the potential for it to save or replace time or work. The technician was attracted by the robot's potential to save time but repelled by the prospect of being replaced. The violinist denied the prospect that robots could make music and thereby replace her. The *My Square Lady* project poured labour into an effort to present Myon theatrically but to avoid gimmickry.

Sarah Thom explicitly commented on her concern about the gimmick aesthetic in my interview with her: 'you can't use everything just cause it's possible. There's got to be a reason for it to be there. Otherwise it's just gimmick.' Thom was clearly concerned here about the potential of the gimmick to irritate. Gob Squad's approach though is bound up with, even defined by, a recognition of the paradoxical attraction and repulsion of novelty, popular culture and new media.

One of Gob Squad's ways of coping with this paradox involves an insistence on the use of newer technologies 'on a DIY level', as Thom put it, so that the audience can see how the effects are being created. Gob Squad's aesthetic engages with and then steps back from the cutting edge through use of anachronistic technologies or by making use of newer technologies in a deliberately home-made or amateurish way. Myon's camp appearance through the floor as comic divinity drew on and drew attention to the way the *deus ex machina* in the theatre now functions as gimmick. It irritates and amuses rather than brings dramatic resolution. It is this comic play with the gimmick's irritating potential that makes Gob Squad's approach so suitable for exploring responses to Myon, and indeed Myon's apparent but failing gimmicky qualities that made it such a suitable temporary Gob Squad member and stand-in for robotics science more broadly. Myon begins to embody ambivalence rather than autonomy (literally, algorithmically, metaphorically and metonymically).

As I have been writing this chapter I have had to continually restrain myself, several times going back and deleting and rewriting the third-person pronoun I have used for talking about Myon. In an attempt to avoid a loaded and gendered anthropomorphism, I have rewritten as 'it'. However, my restrained instinct was always to write 'he'. It is not completely clear why this is the case, although the most obvious reason is that Myon was referred to throughout the project as 'er' i.e. 'he' in German. In a moment towards the start of the performance, Johanna Freiburg playfully drew attention to this:

Johanna: In Kürze werden wir die Vorstellung fortsetzen, sobald sie... er/sie/es [We will shortly continue with the performance, as soon as she... he/she/it]

Manfred: DER Roboter... er [THE robot... he] (here Manfred stresses the masculine form of 'the' associated with robot as a masculine noun)

Johanna: Typisch! [Typical!]

Does *My Square Lady* subtly critique Myon's (en)gendering? Hild comments on Myon's gender as an artefact of the masculine noun 'der Roboter' – a kind of cultural scripting – not of Myon's material design. This design, Hild claims, was deliberately intended to reflect the relatively androgynous physiognomy of an 8-year-old child before differentiation takes place. However, I can't stop thinking that Myon is engendered as a boy. This is a curious reversal given Myon's role as Eliza. An adult woman is played by a (robot) boy. There is, of course, plenty of precedent for this in theatre and opera. There are several subsequent sly references to such reversals in the performance. Christiane Oertel, a mezzo soprano, reflects on her repeated casting in 'Hosenrollen' (so as a male character) and that she never got to play Carmen. Hansky performs Orpheus in the Gluck aria. This is a complex switch in that Orpheus is a male character, but normally in this opera performed by a woman. *My Square Lady* makes use of such gender doubling that is an established part of theatrical traditions – Christiane's long professional career has relied on it. Its choreography sets this tradition in a metonymic relation to the body doubling of the cyborg or robot that was covertly established through the imitation game and makes it explicit.

As I have already referred to, the artifice in artificial intelligence is underpinned by an idea of gendered difference. The imitation game in Turing's paper is played first by an interrogator trying to differentiate a man from a woman before trying to differentiate a human from a computer (Hayles 1999; Turing 1950). The cyborg in cultural criticism and arts practice has offered a potent figure resourcing critiques of power dynamics produced through raced and gendered difference (Parker-Starbuck 2011). For instance, the pop-funk singer Janelle Monáe has mobilised both of these critiques through her figure of Cindy Mayweather, the Afrofuturist Arch-Android who appeared in a number of her albums and videos. For her, the African-American woman was always already doubly cyborg, robot or android, as such imaginary creations have emerged alongside the institutions through which black people were enslaved and women oppressed.[16] This is most explicit in the video of *Many Moons* released on the *Metropolis Suite I* EP which re-enacts a kind of futuristic slave auction (Monáe 2008). Monáe takes hold of this figure to take control of the imaginary at the same time, through her qualified participation in commercial pop music, as acknowledging the impossibility of completely abstracting such technologies from global capital. Her rich aesthetic vocabulary, discussed in much greater detail by scholars of 'Neo-Afrofuturism' Daylanne K. English and Alvin Kim (2013), includes the racial and gender-queer figure of the Android sometimes dressed in a fitted tuxedo, who, in the title of her hit single, precariously walks a political *Tightrope*. As I have already indicated, the female cyborg seems prefigured in Shaw's *Pygmalion*. Eliza for Higgins is a puppet just like Pygmalion's statue. Eliza's rejection of Higgins' claims to her transformation and control (via marriage or other relation of domination) constitutes an act of resistance. Eliza's decision to open a flower shop or marry Freddie exerts her new found if still limited control. Performances by particular actors in this role, such as the one by Mrs Campbell, may even have constituted a resistance to the control of male writers or directors.

In *My Square Lady* something funny happened to this gendered dimension of the story. As mentioned before, the project seemed to offer more of an act of repair rather than a utopian vision of transformation or dystopian warning of overreach. This was not so much a story about the emancipation of women as about the (dis or re)embodiment of men. After all, the imitation game, from Ovid's Pygmalion through to *My Fair Lady* via Alan Turing, is a male fantasy of an autonomous object made alive in the image and imagination of a male creator. When Manfred Hild sang Robbie Williams' and Guy Chambers' lyrics to Myon 'I just wanna feel', there was a certain level of ambiguity as to whether he was singing as himself or about what he wished for Myon. The joke of course emerged from the way Williams' song seemed to speak for Myon. However, the joke works in a particular way when it is a roboticist singing it for his robot. The idea that robots might feel the absence of feeling is an enduring cultural trope – a rehearsal of the Frankenstein narrative. Williams' hugely popular song though seems to have more to do with a crisis of masculinity, a confession of his own (unmanly) depression in tension with desire for (sexual) sensation. In Hild's hands, or rather voice, these readings were doubled. Hild's untrained voice on this stage at this point and his research team as backing dancers foregrounded (white, male) scientists' bodies – and to some extent the weakness of these bodies – stealing the stage from their creation.

One of the researchers, Stefan Bethge, tried to explain the scene to me during rehearsals:

> There's a scene where we first start behind the curtain and then just join in and it's about… actually I don't really know why we're dancing but we're just going to join in from the back and I think it's going to be fun.

His explanation is intriguing in articulating what he does not know and his making sense of this moment as fun. The dancing and the fun seemed to start at the point at which his understanding stops. Since 2008, the American Association for the Advance of Science and *Science* magazine have run the 'Dance your PhD' competition in which scientists are invited to dance their research. Entries tend to involve a choreographic representation of scientific objects but might fruitfully be explored in terms of the way the performances reflect feelings about or relationships with these objects or indeed with the researchers' scientific practice (Tierney 2018).

As the curtain went up on the dancing researchers in *My Square Lady*, there was a whoop in the audience and what seemed like affectionate laughter maybe from family, friends or colleagues. The spectacle of expert scientists singing and dancing inexpertly is a powerful moment of connection. It reconfigures the laboratory as social technology, what Knorr-Cetina (1999) might call 'delaboratorisation', and the stereotypically unfeeling, disembodied scientist as moving, feeling, theatrical body in space.

In normal science, as Stengers (2000b) has observed, the scientists withdraw from the staging of the experiment. The objects speak for themselves and their truth thereby silences fictions. In *Pygmalion* Higgins as feeling male body is repeatedly dragged back into the action by Eliza (and Shaw), however much he might try to keep himself out of it. In performance Mrs Campbell seemed to have performed a similar operation on Shaw himself through her own comic skills. Within contemporary mimetic sciences, reliant on the fiction of the model or simulation, perhaps it is important for the scientist to stand by her or his experiment and this is where Weizenbaum went wrong in his staging. In *My Square Lady*, Hild and his team played the imitation game alongside Myon. They offered their bodies up to question alongside the doubled bodies of the robot, opera singer and performance artists. Katherine Hayles (1999, xii), drawing on Andrew Hodges' (2014) biography, has explained the tragic implications in Turing's own life for his 'being blind to the distinction between saying and doing'. Turing's challenge to gender norms within an oppressive homophobic regime combined with his disembodied understanding of the imitation game with terrible consequences. The collaboration of Hild with Gob Squad and the Komische

Oper might point the way to a reflexively embodied scientific practice: where scientists come out from behind the curtain and the embodied dimension of doing seemingly disembodied science is put on display.

If there was a narrative arc in *My Square Lady*, it was from (the non-events of) Myon's birth to death. Myon's birth at the start of the performance was comically divine, fully grown as *deus ex machina* through a trap door in the floor to be enthroned as Venus on a scallop shell interrupted by Professor Hild. Inevitably the show ended with death(s), although, of course, not Myon's. In fact the whole of the second half of the performance was a series of deaths. It opened with the Antony and the Johnsons' song 'Hope there's someone'.[17] The opera singers walked onto a twilit stage arranged around Myon with three children in the centre. As they sang the ethereally melancholic song, somewhat distanced from the more clearly embodied pain of the original by their technique, children appear and gently touch or hold on to them following the song's plea to 'take care of me when I die'. Later, Orpheus mourned Eurydice's death. Myon was disassembled, did not die, and was nevertheless resurrected or reassembled. Mezzosoprano Christiane Oertel looked back on her long career and, finally getting to sing Carmen, died (as Carmen). Patten and Hild discussed the death drive and astrophysics, Hild confessing how his mortality motivated his work. Patten asked him what would happen to Myon when he dies. The entire cast dropped (fake) dead in front of Myon as Katarina Morfa sang Dido's lament 'When I am laid in earth'. Myon watched them all die and as ever did not do anything to confirm or deny its necropolitical potential. Here though, with bodies strewn across the stage, Myon's twitching attention seemed the most vital. And then, theatrically dying in another sense, it finally sang in a weakly quavering electronic tone: 'I sing the body electric, I celebrate the me yet to come.'

Peggy Phelan (1993, 2003) has argued that the ontology of performance is defined by its disappearance and therefore ultimately by human mortality. *My Square Lady* rehearsed this conclusion repeatedly. The deaths were staged one after another. In response to Phelan, Auslander (2002, 2008) argued that liveness was an artefact of mediatisation. Of course, the deaths are staged in front of or for Myon. Maybe Myon even seemed ultimately to bring them about. *My Square Lady* comically choreographed liveness, indeed life, as both media effect and as performance ontology. It was the experience of both of these apparently contradictory or at least looping logics that produced both nonhuman and human as ambivalent bodies. Singing the body electric staged what subsequently became known as the 'liveness debate' between Phelan and Auslander as musical comedy.

The cast of *My Square Lady* was brought back to (beyond?) life at the end of the performance in a rousing, camp chorus finale. The lyric is from Walt Whitman's poem via the musical number by Michael Gore and Dean Pitchford from the 1980 film of *Fame* about a talented group of young people at a stage school. The sublime notion that the human body can be beautiful in its very materiality and mortality (a kind of materialist transcendence) morphed in the Gore and Pitchford version into the notion that we will all gain immortality by a return to stardust or atoms (another kind of materialist transcendence). In the film the stars were clearly metaphorical – the kids from *Fame* would gain immortality through their mediatised stardom. The sentiment was doubled here. The metaphors all sat beside each other. The sensation was both sublime *and* kitsch. The comic doubling was not simply stating a difference between Myon and the humans on stage in terms of finitude: that the humans would die and Myon would not. Neither was it sincerely staking a new materialist claim for human immortality as energy, atoms or liveliness. Myon's failure to live (in the theatre) was somehow mirrored by a human failure to die (in the theatre). Finitude here, as Zupančič (2008, 52–3) puts it, 'has a hole in it'. It leaks. At the same time, as Harman asserts, objects exhaust our accounts of them. *My Square Lady*'s musical comedy provided a response to the problem of biomimetic dramaturgy and to the experience of the post- or para-human condition. Its ontological choreography resisted the temptation to fill the hole in or settle for a complete account of artificial intelligence either from science or popular culture. Indeed, the comic doubling, to juxtapose some metaphors, created the hole and registered a sense of exhaustion. At the end of the finale number, on screen Myon's past 'life' flashed before its and our eyes and then Myon went onto the street into the and our future. Although we will no doubt be forced to play it repeatedly, as Myon's descendants perform a series of ontological shifts yet to come, staging the imitation game in the theatre is a comic gag. This is the only place where anyone will ever really win.

Notes

1 Shaw wrote *Pygmalion* in 1912. It was first produced and published in German translation in 1913 and then in English in 1914. It has had ongoing popularity and consequent revivals in German as well as English language versions.

2 The *Doppelpass* scheme (2011–21) is administered by the Kulturstifting des Bundes. It is a fund for cultural collaboration supported by the German federal government and European Union regional development funding.

3 This conclusion echoes a brilliant essay by Carl Lavery (2015) that reads laughter at and in Ionesco's work in terms of ecological and environmental politics and in relation to scientific knowledge about climate change.

4 Louise LePage (2016, 283) has critically interrogated the theatre of Ishiguro's Geminoid robots and articulated the way that in his recent robot performances there is a move away from comedy and actually 'a tendency towards sincerity' which seems to constitute an attempt to use theatre to promote acceptance of the technology rather than as a way of exploring different relationships with it.

5 The dramaturgy underpins a much wider range of research in which computers simulate humans or other organisms. The massive Human Brain Project, for instance, is attempting to create an international infrastructure to model the human brain employing hundreds of scientists across Europe.

6 Thanks to Christina Runge, the producer and dramaturg working with Gob Squad on *My Square Lady*, who directed me to this film, which clearly acted as an inspiration for some elements of the project and performance.

7 The scene appears in printed editions but is not always performed. Shaw (1987, 11) stipulates in a note that the scene as written is intended for cinema or particularly well-equipped theatres. It clearly forms the basis of the scene in the 1938 film scripted by Shaw himself and the subsequent musical versions.

8 Certain moments in the first London performance of *Pygmalion* in 1914 were said to have provoked riotous laughter, including the utterance of a shocking expletive that according to the *Telegraph*'s reviewer 'brought down the house' ('Pygmalion, His Majesty's Theatre, 1914, Review' 2014). Theatre historian Fiona Gregory has argued that the comedy was generated by the performances of the star actors playing Eliza and Higgins, Stella (Mrs Patrick) Campbell and Herbert Tree as much as Shaw's dialogue (Gregory 2016). Shaw was ambivalent to some of the hilarity at these performances and indeed to laughter in the theatre more broadly (*Telegraph* 2015).

9 The work done on the project has given rise to a number of postgraduate theses and related publications (Siedel 2015).

10 Critical readings of Turing's original paper on the imitation game have also attempted to restore both Turing's own body and the cultural (gendered) scripts that it rehearses into understandings of computer science and artificial intelligence (Hayles 1999).

11 This is staged in film versions in different ways: as a montage intercut with Eliza's dreams in the 1938 film of *Pygmalion* and intercut with the 'Poor Professor Higgins' musical number in the 1965 film of *My Fair Lady* which includes a comic gag with food.

12 In this chapter I have left quoted dialogue in the original language and provided my translation in square brackets in order to give a sense of the way the code switching between German and English in the performance was part of the aesthetic.

13 In the 1964 film *My Fair Lady*, the assemblage is further reconfigured within the sung numbers when Eliza is assembled from Audrey Hepburn's body and Marni Nixon's voice. In an implicitly new materialist reading of the film, musicologist Marcie Ray detects (a second wave) feminist agency, a cyborg woman in Haraway's terms, emerging from Nixon's voice disconnected from Hepburn's body and Eliza's inner monologue provided by the sung numbers

of a book musical which are also withheld from capture by the gramophone/ Higgins. Higgins cannot record the songs as these take place inside her head (Ray 2014).

14 Dixon (2007) has pointed out that this is a recurring image in the theatrical staging of robots.

15 The idea that objects exhaust accounts of them is developed in philosophical detail in Timothy Morton's influential book *Hyperobjects* (Morton 2013).

16 The word robot was derived by Čapek from an old Slavic word for slave.

17 The trans singer now identifies as Anohni and as such in this song seems to challenge binary understandings of gender as a basis for human relationships.

3

Speculative theatricality: dystopian performatives and vertigo aesthetics in popular theatre

> The scientific version of our existence on this planet may very well be physically true, but we don't like it much. It isn't cuddly. There aren't many tunes you can hum in the shower. (Atwood 2012, 54)

> What is this feeling
> So sudden and new?
> I felt the moment
> I laid eyes on you
> My pulse is rushing
> My head is reeling
> My face is flushing
> What is this feeling?
> Fervid as a flame
> Does it have a name? (Holzman and Schwartz 2006, 146)

The novelist Margaret Atwood snappily sums up a problem driving public engagement with science. Science may satisfy its own obligations towards a particular conception of truth in the production of knowledge. However, such truths seem not to satisfy science's publics in terms of their demands for sensation. Indeed, scientific institutions and individual scientists are often resistant to attempts by popular media to sensationalise even at the same time as they desire engagement. To follow Atwood's line of thought, science often fails to satisfy a particular need to be touched, neither does it offer the reassurance provided by a repeated and repeatable musical form. Atwood, perhaps surprisingly for a writer, seems to conceive of engagement non-discursively as a hug or a melody. Science only makes partial sense. On the other hand, the two female leads in the hit musical *Wicked* play with felt experience through lyrics, gesture, melody, harmony, rhythm and choreography. *What is this feeling?* stages an encounter between two young women that seems to make complete theatrical sense. They have just met each other and experience a set of physical symptoms that musical audiences might quickly recognise as a romantic trope: they are falling in love. At the same time as the staccato recitative of the speculative lyrics quoted above emerges as catchy melody, the speculation is resolved as diagnosis, the feeling is 'loathing, unadulterated loathing': the love song turns out to be a hate song.

In the previous chapter, I examined a set of choreographic practices that held together computer science, opera and contemporary theatre as ways of knowing. In this chapter, I am exploring the way popular theatre might resource practices of knowledge repair. I argue that the theatrical strategies of this kind of performance might hold scientific knowledge and its objects together with public feelings that could otherwise be (mis)conceived as the kind of disengagement, discomfort or antipathy Atwood characterises. Theatre does this through the mobilisation of the conventions which delineate its inside and outside – the cleavages of theatricality in the terms of Josette Féral I discussed in Chapter 1 – and through theatricality's consequent capacity to make things unreal, as I have explored in Chapter 2. Such conventions are mutable as demonstrated by the way a love song can mutate into a hate song but are reliant on an audience's capacity to make sense of them.

The examples I discuss in this chapter, two Broadway musicals and two plays first staged at London's National Theatre, are connected by populist intent and, to varying extents, achieved a level of popularity. Their populism derives in part from particular accommodations with the economies of commercial or partially state-subsidised entertainment. It can also be accounted for by the way they register public feelings or reflect a particular *zeitgeist*. Their various modes of theatricality establish a relationship between performers and spectators but also articulate connections between a range of scientific discourses on climate change, population change, water management, genetics and neuroscience; apparently anti-scientific cultural forms such as myth, mysticism and magic; and aesthetic conventions from literary fiction, pop music and fashion. As such I argue that they constitute a set of speculative theatrical practices.

In their book *Capitalist Sorcery*, Philippe Pignarre and Isabelle Stengers took inspiration from the cry heard at the Seattle protests in 1999: 'another world is possible'. The link that Stengers draws between anti-globalisation movements and her cosmopolitical philosophy of science is rooted in an impulse to speculate about the possibility of another world. With Pignarre she attempts to assemble diverse and sometimes divergent practices to revive a sense of futurity and speculation as a political practice: 'propositions that attempt to transmit effectively – affectively – a feeling for the dangers just as much as the possibilities for us of protecting ourselves from them' (Pignarre and Stengers 2011, 9). Here they echo a number of other thinkers in different fields, from critical pedagogy to queer studies (Giroux 2001; Muñoz 2009), who resist twentieth-century fears of utopian thought in an attempt to conserve the idea of the future, even within the context of the Anthropocene, and escape the otherwise eternal present in which There Is No Alternative to neo-liberalism. This chapter sets out to locate science in performance as just such a set of propositions.

Science is ambivalent towards speculation. Speculative practices are enshrined in some conceptions of science, as in Popper's understanding of a scientific method based on a falsifiable hypothesis that can be tested (2002). The formation of such a hypothesis requires a speculative, imaginative operation. However, scientific literature often positions its findings in opposition to 'mere' speculation. Science's speculative dimension offers great potential for engagement as noted at the start of this book in Stephen Hawking's injunction to be curious. At the same time, its disavowal of its speculative dimension also undermines such an invitation. Speculation is too easily associated with science's Others: intuition, mysticism, myth and metaphysics. Science is often opposed to speculation within the institutional contexts in which it operates, like medicine or law (Chapman 1999; Levin 2016; Sha 2013). Theatre has also shown some ambivalence towards speculation for different reasons. If science's suspicion of speculation stems from the extent to which knowledge might become contaminated by fantasy, theatre's concern seems to be almost the reverse. Playwright Sarah Kane, for instance, opposed speculative theatre to a theatre of sensation or experiential theatre. For her, dramatic traditions that speculate about or in other times and places neglect the experience of the audience in the theatre in the present (Sierz 2002).

I would argue that the practice and conception of science in performance necessarily resists such an opposition between speculation and sensation. Stengers has argued that:

> A speculative possibility does not simply fall from the sky of ideas. Speculation originates in unique situations, which exhibits the possibility of an approach by the very fact that they have already undertaken it. (2011b, 313)

Speculation operates by a kind of catachresis, so a deliberate use of an idea, word or feeling that might not exist in order to enable thought about what does. It tests possibility through a set of techniques or practices available within a particular time and place. Speculative forms though are somewhat paradoxical or oxymoronic – in the way that a term like science fiction is – by invoking both a relationship with reality and a distance from it. Speculative realist philosophy affirms the position that 'cognition grasps a real that is not of its own making, and that its capacities may be reshaped as a function of that real' (Trafford *et al.* 2014, 2).

Speculative theatricality, as a characteristic of science in performance, perhaps risks tautology but is intended to articulate the particular sense of possibility within the unreal that might be felt in the theatre and that I argue constitutes a reparative way of knowing. Within the examples that follow I locate speculative theatricality in particular in the ways theatrical forms hold together sensation and discursive knowledge, so knowledge that can be expressed in language. The speculative practices that emerge from my examples include conventions that bring together scientific ideas, so discourse about climate change, genetics or consciousness, with the tonal, textural and rhythmic possibilities of music, dystopian narrative and mystical or mythical figures from popular cultures. I argue that popular theatrical forms hold political potential insofar as they can gain a hold on ideas because of, not despite, the way they generate raw sensation.

The politics of theatre here is not constituted by the message to be communicated, or by articulating the position of a particular, pre-constituted social class or grouping, nor indeed by a particular mode of production. Rather, this theatre might be considered political through its capacity to enable collective speculation by sensational means in response to the emergence of scientific objects. In the terms of Stengers and Pignarre, such speculation offers a way of dislocating scientific rationality from capitalist reason, concepts which are often equated within narratives of progress.[1] Speculative theatricality achieves such dislocation by expressing popular feelings inspired by such narratives, notably fearful feelings about the future, as they emerge from the promises or predictions of technoscience. It also articulates such feelings with popular cultural figures or scenarios that offer alternative narratives. In *Urinetown* the book musical as cultural form and mode of production holds together the contradictions inherent in the politics of water management. In *Wicked*, the witches provide a doubled figure with which to understand knowledge politics, and the witch-hunt provides a scenario for exploring how public feelings can be implicated and exploited in knowledge production. Discussion of the work by the London-based theatre company Headlong in the second half of the chapter explores theatrical techniques by which sensation and speculation are held together through what I call vertigo aesthetics. In all these examples, non-discursive theatrical elements, notably but not exclusively music, are central as sensational catachresis in holding open space for knowledge repair.

Bad feelings on Broadway: *Wicked* and *Urinetown*

> Lockstock: It's a kind of a mythical place, you understand. A bad place. A place you won't see until Act Two. And then…? Well, let's just say it's filled with symbolism and things like that. (Kotis and Hollmann 2014, 9)

Urinetown is what might almost be called the 'feel-bad' musical. A bizarrely titled show, following a circuitous journey from its author's roots in the alternative theatre scene in Chicago, it opened on Broadway in September 2001 in a run-down theatre in a Manhattan physically and emotionally battered by the attack on the twin towers. It was conceived and written though in the late 1990s and is a direct response by writer Greg Kotis and composer Mark Hollman to 'a fact that trumped all the other facts', namely the conclusion by the scientific community that humans had caused or contributed to potentially cataclysmic climate change (2014, xii). The musical reaches the apparently deeply pessimistic conclusion that the eighteenth-century thinker Malthus was right. Population growth tends to undermine improvements in human quality of life and political systems can do little to mitigate this: progress undermines itself.

In the 2000s, the extreme and urgent facticity of climate change seemed to drive theatre and film-makers towards documentary or didactic modes. Examples might include Al Gore's lecture-documentary *An Inconvenient Truth* (Guggenheim 2006), the 'drama-documentary-animation' *Age of Stupid* (Armstrong 2009), and the pair of performance-lectures at the Royal Court Theatre in London, staged by Katie Mitchell and delivered directly by scientists themselves, *Ten Billion* by Stephen Emmott (2013) and *2071* featuring Chris Rapley (2015).[2] *Urinetown*, however, expresses writer Greg Kotis's 'creeping sense of dread that we're in the process of doing ourselves in' as shabbily spectacular musical theatre (Kotis and Hollmann 2014, xiv). Rather than a rhetorical approach to speculation as in the scientists' lectures, Kotis and Hollman stage an imagined dystopia but also play with the feelings engendered by this scenario and by the book musical as form.

Kotis's bad feelings are given form through a scenario in which years of environmental degradation and drought have left the fictional city in sanitary crisis. Toilet services are run by a private monopoly, the Urine Good Company, in cahoots with a corrupt government and citizens have to 'pay to pee' (Kotis and Hollmann 2014, xii). Tension between the desperate population and those who have to police their toilet use culminates in insurrection. The toilet attendant Bobby Strong leads a revolt against his employers and the conveniences are thrown open to all. Bobby, however, is tricked and killed by the entrepreneur Caldwell B. Cladwell, president and owner of Urine Good Company, who, as his name suggests, earns a good profit from his monopoly and is not going to give it up lightly. Cladwell's daughter Hope, sweetheart of Bobby, takes over the leadership of the rebellion and Cladwell himself meets a sticky end. Unlike in many book musicals though, there is not a simple happy resolution. Cladwell's demise in the revolution does not lead to a rosy future. Hope and the new regime cannot stave off the Malthusian impasse. As Lockstock, police officer and narrator, concludes, 'it wasn't long before the water turned silty, brackish, and then disappeared altogether' (Kotis and Hollmann 2014, 99). It turns out that Cladwell's cruelty was 'effectively' regulating water resources and Hope's well-meant lack of control of the facilities means the city slides back into the stench.

The cultural form of Kotis's 'creeping sense of dread' seems to be the dystopia. *Urinetown* is unusual but not unique in using a dystopian scenario for a book musical. It gains inspiration directly from more obviously historical musicals, notably Brecht and Weill's *Threepenny Opera* and Boublil and Schönberg's *Les Misérables*, but it also makes reference to Sondheim's disturbing fantasies such as *Into the Woods* or *Sweeney Todd*. It displays many of the features of twentieth-century dystopian literary form as for instance outlined by scholars of utopian forms (Cavalcanti 2000; Jameson 2007; Moylan 2000). It is a 'bad place' into which the audience is led by the narrators, Officer Lockstock and Little Sally, who establish its temporal separation from the world of its audience. The narrators do this through their costumes and characterisation recognisable from many cultural representations of mid-twentieth-century urban USA. Lockstock is a gruff, brutal if somewhat comically ineffective law enforcer blending police officers from stage and screen who might include Top Cat's cartoon nemesis Officer Dibble and Officer Krupke from *West Side Story*. Little Sally is a wise-cracking street urchin who seems to resemble stage incarnations of orphan Annie but also nods to characters such as Gavroche from the stage version of *Les Misérables* or the Artful Dodger from Lionel Bart's *Oliver!* Together they offer a narrative commentary that both establishes the dystopian fictional context of *Urinetown* but also establishes its credentials as homage to and parody of the book musical, as in the quotation at the top of this section. Urinetown is recognisably a US city – it could be New York or Chicago – but its 'Gotham-like' set and costume design, along with the frequent nods to Brecht and Weill, perform a time-shift, a dislocation, that is both backwards into the Great Depression of the 1920s or 1930s and sideways into an alternative world after 'the Stink Years... when the water table started to drop and then just kept on dropping' (Kotis and Hollmann 2014, 78).

The paradigmatic dystopian novels of the twentieth century – Zamyatin's *We* (1993), Huxley's *Brave New World* (2007) and Orwell's *1984* (2004) – and their theatrical equivalents such as Čapek's *R.U.R.* marked their separation from their contemporary societies by a time-shift into the future signified, at least in part, by the profusion of sophisticated technologies of surveillance and control. They warn against the potential of these technologies to support the emergence of totalitarian regimes. *Urinetown*'s technological focus is the toilet. Suitably, in an age of biopolitics, this dystopia is viscerally of the body and the humour is scatological. The technology itself is of course not speculative. In parts of the world where *Urinetown* has largely been performed this is a rather taken-for-granted piece of technology. However, it points to the way that the dystopian potential of technology is not always immediately apparent and that progress can no longer depend on unlimited natural resources. Changing water levels and patterns of precipitation causing drought and flooding are all potential and current effects of global warming. Recent histories of water availability and use may not be a reliable guide to future hydro geographies. *Urinetown*'s mode of speculation therefore has a complex temporality. It looks back, through its twentieth-century milieu and homage to the book musical, in order to look forward speculating on a future ecological crisis.

The Marxist literary critic Fredric Jameson has criticised the anti-utopian tendency within twentieth-century, particularly Cold War, dystopias, notably Orwell's *1984*. Jameson draws attention to the 'ecumenical' negativity of these works, emerging from a fear of the future and progress as such, which seems to foreclose on any sense of political possibility. This critique is based in part on a logical contradiction he identifies in *1984* between the idea that totalitarian regimes can exercise power through science at the same time as being fundamentally in opposition to it (2007, 200). *Urinetown* seems similarly ecumenical in its political critique. The invidious choice seems to be between a form of capitalist totalitarianism in which the majority suffer and corporations and state connive to profit from regulation regimes, or anarchic freedom ultimately dooming everyone. It is not so much the sanitation technology that is put into question, but its availability and governance. Science is at a standstill and no mode of government can equitably regulate access to technology.

And yet narrative analysis eventually runs dry when trying to understand dystopia in performance. It rather misses the fact that much theatre, the musical in particular and entertainment in general, are not just capable of formally representing utopia but of temporarily constituting it in performance (Dolan 2005; Dyer 1992). While the dystopian scenario on stage may seem to reflect public fears about climate change or overpopulation, rather than the Cold War fear of totalitarianism or nuclear apocalypse, the feeling in the audience is unlikely to be one of overwhelming anxiety. The relocation of the dystopia is dislocated by the theatrical event.

Urinetown closely follows all the conventions of the book musical and as such has an emotional trajectory at odds with the creeping dread that was Kotis's starting point. There is a (hetero)romantic story-line between Hope and Bobby Strong that is consummated in a love song *Follow your heart*. This is classically in waltz time, starting as a solo for Hope and turning into a duet with Bobby. The singing is straight (in both senses) rather than comic (or queer). Jennifer Laura Thompson in role as Hope in the Broadway production makes a marked shift between a comic, girlish speaking voice and a classically musical, technically accomplished sung soprano. The gospel number *Run, Freedom, Run* at the start of Act 2 is suitably uplifting and the final number *I See a River* moves from gospel to a hymn marching hopefully in solidarity into the future.

Critical literature on the dystopia makes a distinction between the extent to which escape or utopian potential exists within or without the fictional world. Moylan and Baccolini identify a key feature of the (particularly feminist) critical dystopia as the preservation of hope for protagonists within the world and readers without through 'ambiguous open endings that resist closure' (Moylan and Baccolini 2011, 14). The narrative ending of *Urinetown* seems to offer complete closure: it does not offer redemption through love, individual self-preservation or collective emancipation. Hope and Bobby's love is naive and clichéd. Bobby is killed and Hope's revolution comes to nought. *Urinetown* is self-consciously Brechtian in form and politics in that it articulates contradictions in the mode of production and prompts laughter at an apparent state of misery. However, the audience is as likely to leave the theatre humming the tunes as planning climate activism:

> Lockstock: Like I said, Little Sally, this isn't a happy musical.
> Little Sally: But the music's so happy! (Kotis and Hollmann 2014, 99)

The musical experience, given that it is so closely aligned with other book musicals, is likely to produce similar affects. Audiences are not supposed to go to Broadway musicals nor leave successful ones in a state of anxiety about the world. And this is perhaps where the utopian potential might be found, to echo Jill Dolan (2005), through what might be called a dystopian performative. If the dystopia enables us to feel for the danger and feel the fear, its performative dimension lies in what it does.

As Anne Beggs (2010) has pointed out, the particular twenty-first-century twist on the book musical performed by *Urinetown*, is its constant reference to itself as musical theatre and the extent to which it is, is not or should be observing the conventions of the form.[3] This runs right through its formal properties into its own mode of production. Little Sally comments that there must be hope that her dreams will come true, purely because a little girl like her has been given as many lines as she has in a musical. In an apparent analogy with the inevitability of environmental degradation, *Urinetown* reflects on the inevitability built into its own theatrical form. Just as the musical proceeds inevitably through its conventions, even though the characters are aware of them this does not make them able to stop them. The audience is also interpellated in this way. Except that Beggs identifies a number of respects in which *Urinetown* demonstrates a sense of possibility:

> For *Urinetown*'s real subversive potential lies not in its obvious critique of capitalism, but in its paradoxical refusal to carry immanent musical conventions through to their anticipated conclusion: the songs are tuneful and the love story touching, but they do not solve manmade environmental crises or their concomitant political exploitation. (2010, 55)

In this way she suggests *Urinetown* makes and then subverts a particular contract with its audience and that, in its rise from fringe parody to Broadway hit, models a kind of political subversion within the mainstream and the 'middle-brow'. She goes on to show how the musical's reach has been enhanced by touring and revival hundreds of times across amateur and student performances.

On the one hand, *Urinetown*'s commercial success and commodification in all the usual merchandise spin-offs – t-shirts, CDs, etc. – can be understood as inevitable recuperation. On the other, *Urinetown* has already incorporated this recuperation within the DIY aesthetic. The cast recording sleeve note is introduced with the prefix that it is 'produced by a large monopolizing corporation'. Now this kind of reflexive *schadenfreude* might be dismissed as cynical or anti-utopian in Jameson's terms. However, I think this is to disregard the significance of its potential to produce an alternative public that might cut across established lines of identification. Generic forms can become what Baccolini in her work on the feminist critical dystopia has referred to as 'sites of resistance' (2004, 520). The symbolic inability of the protagonists within the world to figure a way out of the dystopian bind that is *Urinetown* is materialised in the inability of the musical's creators to escape the economics of Broadway. At the same time, the audience's collective recognition of this impasse, indeed their potential irritation at it, at the same time as they enjoy the catchy melodies, suggests at least the possibility of the dystopian performative having affects or effects beyond the sense of no alternative.

Pignarre and Stengers identify a danger in assuming that the scale of ecological catastrophe in itself can force consensus and unity. The widespread documentary and didactic approaches that characterise much science communication, including the documentaries and performance lectures mentioned above, tend to focus on delineating this scale and emphasising its facticity, attempting to iron out the apparent differences that are understood as disagreement about the facts of climate change (Hulme 2009). However, this does not seem to be enough, neither does it make a virtue of theatrical convention. Perhaps *Urinetown*'s more circuitous reflexive theatricalities offer a greater sense of possibility in terms of creating 'practices that may turn these divergences into a force' (Pignarre and Stengers 2011, 9). At the very least, it might upset the opposition posed by Atwood between the sensation of humming tunes in the shower (or on the toilet?) and scientific speculation.

If *Urinetown* became a surprise Broadway hit of the new millennium with an unlikely dystopian scenario, then an equally surprising hit two years later was built on the figure of a green witch. *Wicked the Musical* (2003), written by Stephen Schwartz and Winnie Holzman, shares with *Urinetown* many features, notably a reflexivity in relation to its own form and the dystopian depiction of an onstage public. This depiction resonates with the way the public is (mis)understood within its deficient conception by the public understanding of science movement: that is fearful, gullible, ignorant, prejudiced and quick to scapegoat. It is also similar and operating at considerably greater scale in producing an off-stage public through reflexivity towards contemporary knowledge politics. Its reclamation of the figure of the witch, importantly two witches in this case, is central to this variant of the dystopian performative.

The specific witches in question here are the double leads in *Wicked* and the opening quotation of this chapter is from a duet they sing which paradoxically both establishes their friendship and their loathing for each other. *Wicked the Musical* is an iteration in a long-running story at the heart of US culture. It is a musical based on a 1995 novel by Gregory Maguire, which adapts the 1939 movie *Wizard of Oz*, itself based on the 1900 novel by L. Frank Baum (2019).[4] The story in its numerous incarnations in novel, stage or film relies on a journey by a young female central character into a utopian or perhaps dystopian fantasy world of Oz. The twist in *Wicked* is that the central character is not the Kansas every-girl Dorothy but the Wicked Witch of the West who turns out to be not so wicked (at least in the traditional sense) and an every-girl for the twenty-first century. The musical focuses primarily on the intertwined back stories of Elphaba, the Wicked Witch of the West, and Glinda, the Good Witch.[5] In many ways these two witches are two halves of the same character. Scholarship on the musical has shown how their complementarity is played out within the form through their duets, through the marketing logo, through their growing friendship despite their initial apparent incompatibility and through the way the two characters and the actors who play them generate powerful attachments with their many young female fans (Boyd 2010; Wolf 2007, 2008).

The musical, as is typical of the classic book musical form, has a dual trajectory through the sung numbers and through the book narrative. The sung numbers recount a romantic journey and a process of self-realisation. Both these journeys are primarily for the two female leads. Stacy Wolf (2008) outlines the significance of *Wicked* both in its popular feminism but also in the way that it queers the central romance. As Wolf points out, the male lead and love interest of both girls, Fiyero, is a 'straw man' in both senses and the real romance is the friendship between Elphaba and Glinda. The book narrative tells the story of Glinda and Elphaba's education at Shiz university, their discovery of the persecution of the Animal professors as part of an apparently fascist tyranny over Oz orchestrated by the wizard and the teacher Madame Morrible. Elphaba and Glinda are both concerned by this but adopt very different approaches. Elphaba directly confronts the wizard, speaking truth to power and counterposing his apparently illusory use of technology to inspire fear – through a rather clunky and clearly early twentieth-century mechanical head – with her real power stemming from her ability to read the Grimmerie and therefore to do magic. Hers is an organic, biological, reproductive power to transform animals and people and a power to fly and make things fly. Like the normally male figure of the over-reaching scientist she struggles to control this power and is prone to a degree of hubris.[6] Her 'curing' of her disabled sister Nessarose is politically problematic, in that disability seems to be somehow connected to malign behaviour. Elphaba is also complicit in her sister Nessarose's death, which results, as at the start of the 1939 film, by Dorothy's house squashing her flat. We do kind of understand how she becomes wicked in the eyes of the public even though this is also obviously simplistic and unjust.

If wickedness in the 1900 novel was clearly a moral concept, a century later it seems to have mutated somehow. The villain a century ago now turns out to be the heroine, however flawed. The source of moral authority, the good witch Glinda, is now a vacuous, if well-meaning, politician. Elphaba's wickedness, and indeed wickedness more broadly in the musical, derives from apparently uncomfortable knowledge. Elphaba's knowledge is uncomfortable in the sense that it emerges from her arcane, magical abilities that she can scarcely control, but is also reflected in her awareness of what the wizard is really doing that everyone else is either unaware of or chooses to ignore. This situation clearly resonates with the teenage fans of the musical who may be coming to terms with new but frightening problems and knowledges of adulthood. However, wickedness understood in this way is also a conception of contemporary problems co-produced by science and society. 'Wicked' problems are not just problems that produce disagreement about how to solve them; they are also subject to contestation about how they are conceived or framed in the first place (Rittel and Webber 1973). Scientific institutions, by definition, struggle with wicked problems, as the knowledge that defines them as discipline or profession is insufficient to address the problem. At the same time, acceptance of other knowledges may fundamentally disrupt or threaten the institution or social group. Steve Rayner argues that ignorance is often a 'necessary social achievement' in shoring up institutions (Rayner 2012, 108). He goes on to theorise 'uncomfortable knowledge' as being at the meeting point between 'wicked' problems and 'clumsy solutions' that might take into account different ways of knowing.

Discourses about scientific over-reach, in the theatre or elsewhere, tend to either focus on the moral dilemma faced by individual scientists or direct moral critique at the way particular social systems determine scientific knowledge production and, in particular, use it.[7] However, the tendency in these accounts is for the knowledge itself to be unproblematised as knowledge. Wickedness in the musical registers the loss or indeed absence of a safe moral ground on which to make judgements about knowledge. It also clearly articulates the way knowledges might be disqualified, ignored or rejected both through authoritarian repression but also through more hegemonic psychosocial operations by experts and non-experts. The book narrative of *Wicked* relies on a set of historically significant figures and scenarios for the exploration of uncomfortable knowledges and public ignorance in its featuring of the witch, witchcraft and the witch-hunt.

The witch-hunt as staged in *Wicked* and indeed elsewhere in culture is often read as metaphor for the demonisation of a minority group. However, such metaphorical usage risks further suppression of its historical actuality. Silvia Federici (2014) has shown how, in the sixteenth and seventeenth centuries across Europe and in various colonial territories (notably North America), the witch and her magic were produced as objects for popular fears about social change. Such changes included the land enclosures, eradication of the commons and associated capitalisation necessary for a phase described in Marxist terms as 'primitive accumulation'. The witch-hunt is an enduring mode of public engagement whereby a weak or weakened group is interpellated as an object for public feelings that might otherwise be object-less or at least where the object(ive) of the feelings is elusive. As modern science accumulates the power to produce new objects, its emergent institutions inevitably become caught up with such modes of engagement particularly around questions of epistemology. In her powerful critique of modern science, Carolyn Merchant (2006) showed how the proto-scientific utopian imagination in the seventeenth century, as exemplified in the work of Francis Bacon, was permeated with the vocabulary and imagery of the (then contemporary) witch-hunt.[8] The witch-hunt both secured and resourced the interconnected emergence of modern science and capitalism. *Wicked*'s twisting of the *Wizard of Oz*, in a way that pulls critical focus onto the witch-hunt as dystopian scenario, offers potential insight into the way modern science has historically positioned its publics.

Stengers (2011a, 375) has criticised the apparent need for modern science to eliminate other knowledge-making practices and associated this with the tendency for scientists to adopt 'the role of guardians of an infantile public'. In the *Wizard of Oz* and *Wicked*, the wizard is interposed as the figure of the 'quack or swindler' to whom the gullible public is likely to fall victim. At the start of the twentieth century in the *Wizard of Oz* he appears a fairly benign avuncular figure. A century later in *Wicked*, he still professes to be 'One of your dime-a-dozen, Mediocrities' who 'got carried away' in a musical number *Wonderful* that, with its ragtime piano, musically evokes the Edwardian era (Holzman and Schwartz 2006, 167). At the end of the number it becomes a duet with Elphaba, although she discovers the full extent of the wizard's oppression of the Animals: a somewhat condensed analogy for the collapse of the progress narrative.[9] The wizard sings, 'The truth is not a thing of fact or reason / The truth is just what ev'ryone agrees on' – a line written prior to the election of Trump and the popular debates about post-factuality. For Elphaba and Glinda, this is posed as problem rather than accepted condition. For Elphaba this means that she will always be opposed to him.

The doubled figure of the witches Elphaba and Glinda is appropriate to exploration of (scientific) knowledge's promises and the mixed up feelings generated by them. According to Stengers and Pignarre, the contemporary (neo-pagan) witch is a political figure who challenges the way modern science might rely on the disqualification of particular indigenous or women's knowledges. Its reference to the historical context discussed above also forces an examination of the complicity of certain scientific institutions with capitalist reason, or what they call *Capitalist Sorcery* in their book title. For them this move is pragmatic in challenging oppositions between (rational) knowledge and its (superstitious) others including easy oppositions between knowledge and belief: 'the very naming of capitalism as sorcery transforms the relation with those "others" supposed to be superstitious' (Pignarre and Stengers 2011, 40). Similarly the reclamation of the figure of the witch is a creative political act intended to reanimate, or at least open up a space for, lost or disqualified knowledge.

If the promises of the patriarchal figure of the wizard, who in the musical turns out actually to be Elphaba's father, are clearly a malign con, Glinda and Elphaba's knowledge of witchcraft is also itself uncomfortable and poses a political dilemma within the dystopian context or indeed within the context of patriarchal modernity. In terms of speculative (object-oriented) political philosophy, this dilemma might be characterised in Graham Harman's terms (2014, 3) as that of truth vs. power politics rather than that of left vs. right. Elphaba's truth politics might refer to her whistle-blowing in the face of the wizard and Morrible's manipulations or indeed to the US cultural trope of being true to oneself. It is not that Elphaba, even with her access to magic, has unmediated access to an underlying truth, but at moments she thinks she has and it is on this thought that she bases her actions. Glinda's power politics emerge from her charm and popularity and an apparent lack of a need for any other ethical framework. As she sings in her defining number:

> Celebrated heads of state or
> 'specially great communicators
> Did they have brains of knowledge?
> Don't make me laugh!
> They were popular! Please-
> It's all about popular! (Holzman and Schwartz 2006, 152)

While remaining in sympathy with her friend, she takes on the role of Governor working with the wizard and organises the hunt for the exiled Elphaba. By the end of the show they both come to a greater understanding of each other's philosophies without being able to reconcile them. Despite her continuing revolutionary attitude, Elphaba acknowledges in song 'I'm limited' revising her earlier defiance of limitations including gravity (Holzman and Schwartz 2006, 175). In the same duet, Glinda reflects on her lack of access to magic – and the ability to search for truth – via the Grimmerie. All she knows and what they both agree on is that they have been changed by their relationships with each other. The similes in their duet are object oriented. The real reference point for their subject-subject relations is in object-object relations and in that sense is unfathomable. This is not resignation – the promise of change is kept open and Glinda promises to learn. However, it is a recognition of limitations in the possibilities for reconciling contradictions between truth and power.

Glinda:	Elphaba:
Like a comet pulled from orbit	Like a ship blown off its mooring
As it passes a sun	By a wind off the sea
Like a stream that meets a	Like a seed dropped by
Boulder half-way	a bird in the wood.
Through the wood.	

Both: Who can say if I've been changed for the better?

(Holzman and Schwartz 2006, 175)

Oz in *Wicked* is utopian (like Oz in the *Wizard of Oz*) in that it is a realm in which young women have agency through their acquisition of knowledge and through access to power. Like the *Wizard of Oz* which has made a somewhat contested contribution to queer utopias, *Wicked* also seems to offer a kind of utopian performativity that extends well beyond the experience of the show in the ways that young women seem to use the songs, the story and the characters in developing their individual identities and relationships with their friends (Cavendish 2016; Dvoskin 2011; Wolf 2007). However, this is achieved within a dystopian environment, one in which regimes of progress and modernity seem to run counter to the women's solidarity, friendship and self-realisation. Importantly also the exercise of power seems to be at the expense of nonhumans in the world. Elphaba's resistance is triggered by her concern for the maltreatment of the goat professor and the lion cub. The ending and indeed the tone of the musical are bitter-sweet. Although Elphaba escapes and gets to be with her straw man Fiyero, it comes at a great cost: she loses Glinda who is not even aware that she is still alive. While Elphaba and Glinda display a certain resilience in surviving and there is a promise of a better future for Oz under Glinda's government, Elphaba and Glinda can still not be together.

The power of the witch-hunt embodied in the ignorant on-stage public, easily manipulated and prone to scape-goating, still somehow holds sway. However, the musical does not position off-stage publics, i.e. the theatre audience, in this way. *Wicked* engages because it provides moments of joy and self-realisation at the same time as registering anxieties about truth, knowledge, power, solidarity and relationships with the nonhuman world. Glinda's and Elphaba's knowledge may be incompatible in some senses; however, they are bound together within a musical romance.

The doubled figure of the witches offers an object of attachment for young people to speculate about the possibilities of friendship within post-factual society. Similarly the musical also holds together but holds apart ways of knowing constituted by the book, the music and its cultural reference points. That storytelling conventions, such as the musical, exclude or leave out knowledge is accepted as necessary to the form and is often deeply understood by its audiences. The way *Wicked* holds together Glinda's and Elphaba's apparently contradictory rationalities, gestures towards a much wider political potential for modes of speculative theatricality.

Both *Wicked* and *Urinetown* turn back nostalgically to historical contexts and book musical conventions. They rely on their audiences' reflexive awareness of these moves for many of their affects and effects. It is this collective nostalgia, which as Baccolini has asserted is often regarded as conservative, that constitutes the audiences as publics within and beyond the theatres and sets up the mode of theatrical speculation. Baccolini (2011, 175) argues that it is the 'combination of desire and estrangement and the presence of a slight suffering' that can make nostalgia a resource within feminist critical dystopias. This combination resides here in the theatricality of the nostalgia – the doubleness (existence and non-existence) of the place that has been lost in the past. Both shows recognise this doubleness through the reverent and yet paradoxically satirical relationship that both shows have to their form. It is also this relationship to their convention that resources their utopian performatives, in Dolan's terms, such as the new orientations towards knowledge politics arrived at by the witches at the end of *Wicked*.

Falling headlong: vertigo aesthetics

The on-stage worlds of *Wicked* and *Urinetown* do not take place here and now even while, as I have shown, they clearly resource certain utopian impulses in their contemporary audiences. *Wicked* and *Urinetown* reflect speculative theatricality within the tradition of the commercial Broadway book musical. However, neither a complete fantastical shift in space or time, nor the book musical form is a necessary feature of the dystopian performative or speculative theatricality more broadly. A dystopian brand of speculative theatricality can also be found in twenty-first-century British theatre. Merle Tönnies has argued that a number of twenty-first-century British plays – she focusses on work by Edward Bond and Mark Ravenhill – use but warp the form of the (classical) dystopia. She claims that the significant departure from dystopian form articulates a sense of stasis or 'immobility' both in character onstage and audiences off. The apparent possibility for change is absented leaving behind what she calls a 'haunting quality' (2017, 169). In a similar vein, Dan Rebellato (2017) has remarked on an 'anti-realist apocalyptic tone' in British theatre at the end of the twentieth and start of the twenty-first century. He identifies the apocalyptic tone across British writers' theatre including Sarah Kane, Caryl Churchill, Alistair Macdowell and Alice Birch. Rebellato comes to the idea of apocalypse through noting the repeated use of violent (biblically) apocalyptic imagery and 'a hypertrophic escalation in the scale of violence represented'.

Along with both these scholars, I detect a somewhat elusive quality or tone that articulates a feeling of impending doom or chronic despair and characterises a particular form of science in performance. Rebellato refers to the anti-realism of his examples in representational terms, as for him, following Mark Fisher (2009), realism is redundant under neo-liberal capitalism. It is interesting that, in his use of the apocalypse rather than the dystopia as a formal reference point, he also seems to suggest they reach away from a modern or scientific worldview towards a quasi-religious revelatory imaginary to dislocate the contemporary feeling. Much of the work though articulates with the recognisable objects of speculation discussed in this chapter: climate change (e.g. Churchill's *Far Away* (2003)) and the psychopharmacological revolution (e.g. Kane's *4.48 Psychosis* (2001)).

The London-based theatre company Headlong has a track record of dystopian visioning that is clearly influenced by the work that Tönnies and Rebellato discuss. They also occupy a similar position in UK theatre ecologies in London's producing houses for new writing. However, the self-styled 'company for the digital age', founded in 2005, has developed a distinctive mode of speculative theatricality that resonates with the dystopian performatives of the musicals discussed above in articulating speculations about science's objects and ambivalent public feelings (Headlong n.d.). In Headlong's work the legacy of an earlier generation of writers such as Sarah Kane, memorably associated with 'in yer face' theatre (Sierz 2002), is detectable in bold staging and a rich sensory audience experience. However, Headlong's generation of playwrights – all born around 1980, so starting their careers almost exactly at the millennium – that might include Mike Bartlett, Duncan Macmillan, Lucy Prebble and Ben Power also seem to be influenced by documentary or political theatre traditions. Headlong's work in its first decade therefore was socio-politically engaged but without a clear ideological perspective. It attracted a popular audience and combined a somewhat dark view of contemporary society with a grasp of hedonistic popular culture.

The company's mode of speculative theatricality might have been set consciously or unconsciously by Rupert Goold's first production as artistic director, which was an adaptation by Ben Power of *Paradise Lost* (2004/6).[10] Headlong's programming has subsequently included contemporary stagings of classic dystopias such as *1984* (2013) in an adaptation by Robert Icke and Duncan Macmillan and continued under Goold's successor, Jeremy Herrin, with new plays such as *The Nether* (2014) by Jennifer Haley and *People, Places and Things* (2015) by Duncan Macmillan.[11] Headlong does not present itself as a theatre company that explicitly sets out to work between science and theatre. The company had a pre-Goold history of touring modern classics in its previous incarnation as the Oxford Stage Company. This heritage continued to be detectable in its programming under Goold and probably contributed to close collaboration with the UK's mainstream subsidised theatre establishment including London-based institutions like the National Theatre and regional producing theatres. Its programming as Headlong combined new plays by emerging and mid-career writers and adaptations of classics. But all their work displayed a consistent concern with capturing a *zeitgeist* through an aesthetic under Rupert Goold – what I am calling their mode of speculative theatricality – that became something of a trademark.

A trailer for the company's 2012–13 season, directed by Goold in his last full season as artistic director and entitled *Falling Headlong*, synthesised the feeling of the five works of that season (*The Effect*, *American Psycho*, *Boys*, *Chimerica* and *Medea*). This trailer was originally released on the Facebook social media platform and at the time was regarded as a pioneering piece of marketing (Alberge 2012). It was specifically intended to attract younger audiences than the National Theatre's average demographic. Set in an apparently vacated office space, individual characters from these pieces are pictured alone, lost, confused. There are images of objects, floating and falling in space, people rushing in slow motion down empty corridors, walls closing in. The backing track by James Blake repeats the verse 'All that I know is I'm fallin', fallin', fallin', fallin' in an aching voice against an ethereally melancholic electronic soundscape. The music critic and theorist Mark Fisher wrote about listening to Blake's musical trajectory as 'like hearing a ghost gradually assume material form' (Fisher 2013a). This ghost and the melancholic tone of Blake's music reflect what Fisher calls a 'secret sadness' within twenty-first-century hedonistic (techno) culture. It is an apt soundtrack to Headlong's season and much of their work under Goold and beyond.

A preoccupation of the company seemed to be this sense of vertigo, running or falling headlong unable to stop. This feeling is materialised in different ways within the often loosely realist – at least in terms of being clearly located within a contemporary socio-historical setting – worlds of the plays. Technoscience is heavily implicated in this fall from grace. The failings or limitations of psychoactive medication, administered by ourselves or others, as a coping mechanism is thematised in Lucy Prebble's *The Effect* (2012) and Duncan Macmillan's *People, Places and Things* (2016). The morally and politically destabilising effects of digital technologies provide the object of anxiety in *The Nether* (2015), the 2014 digital installation *Prism* and their versions of *1984* (2013), *Spring Awakening* (2014) and *Pygmalion* (2017).

Mike Bartlett's play *Earthquakes in London* (2010) directed by Goold and designed by Miriam Buether engages with the scenario of impending ecological devastation that recurs throughout this chapter.[12] It is largely set in a recognisably contemporary London but starts in 1969 and finishes in the year 2525. The 1969 song by Zager and Evans that plays at the start and end of the piece, asking whether humanity will last to 2525, sets up the premise for the play. The dystopian proposition of the play sees the (long) present since 1969 as the end of the future. The post-apocalyptic fantasy (set in 2525 or a contemporary hospital) at the end of the play perhaps presents a glimmer of a possible post-apocalyptic utopia to come, although this could in fact be the dying hallucinations of the central character Freya who jumps or falls off Waterloo Bridge. Freya is one of three daughters of a now estranged climate scientist father Robert, whose youthful optimism and commitment to science and the planet in 1969 is sold out in the 1970s to a multinational airline and by the present (of the play) has become an embittered fatalist.[13] Robert is neither a Frankenstein directly responsible for the planet's doom, nor a heroic figure that has not been listened to. He is rather a deeply flawed cynic whose views, that the tipping point has passed and that the earth will long outlive humans whatever damage has been done, appear to echo those of the iconoclastic James Lovelock (Bottoms 2012). He coldly advises his pregnant daughter Freya not to have her baby.

This sense of doom-laden inevitability and disempowerment is resonant with *Urinetown* and, in a similarly theatrical move, the apparently depressing conclusion of the play was mitigated in its original production by its fast-paced, theatrically rich staging. In Buether's design there was a catwalk stage that snaked through the auditorium, letterbox openings at either end of the space and projections onto the walls. The published script directs producers to stage the play with 'as much set, props and costumes as possible' (Bartlett 2010, 5). The impression though in Buether's design was not so much of lots of stuff. Because of the narrowness of the catwalk and the letterbox stages which featured flashback sequences in the action, it was easily crowded. Rather costumes and props proliferate in ways that are both banal and magical. They rapidly appear and disappear, as in Marx and Engel's articulation of the sorcerous ways that capitalism constantly changes its means of production, 'all that is solid melts into air' (Marx and Engels 2015, 11). In one of the bold musical sequences, identikit, sinisterly glamorous, black-clad mothers in dark glasses (think gothic Audrey Hepburn) take their babies from their prams and throw them into the air – they disappear into black confetti. Scenes take place alongside and over the top of each other backed by a series of anthemic popular music tracks of the first decade of the twenty-first century. There are moments where the choreography and the the atrical environment play with the precarity of the characters, flirting with apparently real risk for the actors. In a post-show discussion, Goold recounted an anecdote about an accident on press night when one of the actors fell down a trapdoor and got lost below the theatre trying to get back on stage.

In this scene in Act 4, Freya discharges herself from the hospital where she has been admitted in a desperate dilemma about her pregnancy brought on by the conversation with her father. She encounters an Eastern European cleaner, Maryna, who in a mode of musical theatricality lip synchs to the Marina and the Diamonds (2009) track 'I am not a robot', a confession of vulnerability that Freya cannot seem to get anyone to hear. In a highly complex choreography, they dance through and with a series of characters that pass by on the narrow stage. A hospital porter pushes a woman in a wheelchair; patients on chairs pick them up and leave; commuters on mobile phones take the cleaner's costume as she transforms into the Marina of the hit song in a tight pvc dress; a bike cycles through just before road-workers come up through a trapdoor; a thief is pursued by the police; Freya takes a picture of some lost tourists; a street performer poses in a silver lamé costume; a boy breakdances; an underground train passes with commuters strap-hanging; and Freya is lifted into air by the commuters (Figure 3.1).

The everyday experience of walking down the streets of Central London is rendered mythic as technicolour pop video. Freya is buffeted bewildered, ultimately fatally for her and, one assumes as the earthquake of the title shakes the city, for much of the population of the country if not the planet. The audience in the Cottesloe performances at the National Theatre presumably came down these very streets on their way to the theatre but now get to sit amid the spectacle, at eye level for those in the stalls, as the commuters rush by oblivious to the coming apocalypse.[14] If the audiences of *Urinetown* are interpellated as (potential) co-producers and *Wicked*'s girl fans as friends of the witches, here we are spectators of our own doom, emancipated maybe, for all that it's worth.

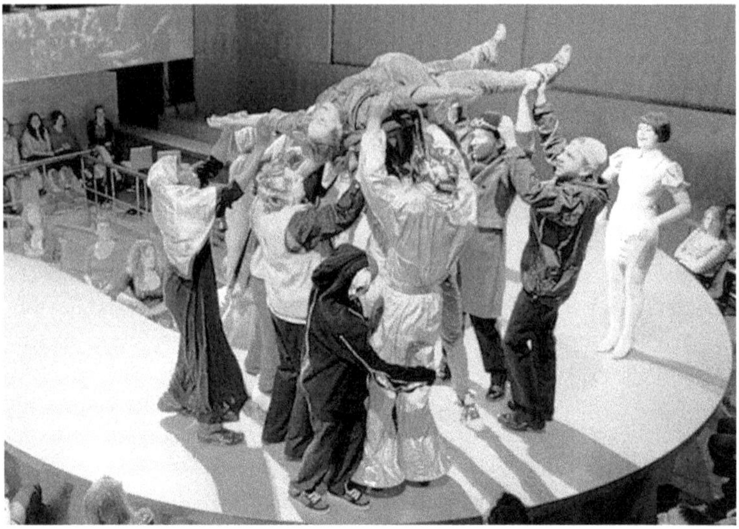

Figure 3.1 *Earthquakes in London*

Here though the only possible utopian community is in the theatre where, for a moment (well nearly three hours), you get to pause and reflect on the headlong fall.

The sense of excess that is expressed literally through the profusion of theatrical resources devoted to staging *Earthquakes in London* is stripped back significantly in Lucy Prebble's *The Effect* staged two years later in the same theatre. Here the five-act epic with a crowd of characters is shrunk to a tighter structure involving two pairs of actors: two women and two men playing two clinical scientists and two patients. The play follows a clinical trial of a new anti-depressant. Connie and Tristan, played in the original production by charismatic duo Billie Piper and Jonjo O'Neill, are young adult volunteers testing the drug under the supervision of Dr Lorna James and Dr Toby Sealey. The audience sits on four sides and are witness to an experiment that goes badly wrong due to the manipulations of both patients and doctors. As they are given progressively bigger doses Connie and Tristan start to experience symptoms of heightened sensation:

> Connie: I've never felt this alert before.
>
> ...
>
> Tristan: I feel. Full. I feel almost... holy. Like life is paying attention to me. (Prebble 2012, 52–3)

The Effect's scenario, in its representation of apparently contemporary scientific practices with no time-shift or obviously fantastical displacement of the action, is less obviously dystopian than the other examples in this chapter. Perhaps for this reason, it has been the subject of more extensive plausibility testing within (scientifically or clinically) expert reviews of the play (Boyce 2012; Zuger 2016). Its proposition for Connie and Tristan, that might be read as dystopian, is that falling in love may lose its meaning because neither can determine if the feeling is real or the effect of a psychoactive molecule in the drugs they have been given; or because real love or happiness just *is* the effect of a psychoactive molecule. At the end of the play, having suffered a near fatal overdose, Tristan has serious amnesia and is a shadow of the wittily charming character at the start. The doctor, Lorna James, who it emerges has long suffered from depression and refused to take medication, is stuck in bed, her body unable to cope with her sense of blame for the calamitous ending of the drug trial: 'I don't have enough skin' she says (Prebble 2012, 98).

The position here is not in the arguably utopian traditions of critical or anti-psychiatry that are less ambiguous in their criticism of the psychopharmacological revolution and dominance of drug treatments (and Big Pharma) in the treatment of mental distress. Such ideas are voiced in the play notably by Lorna, but they are juxtaposed with apparently contradictory plot features and the contrasting perspectives of other characters. In a long argument between her and Toby which stages starkly opposing positions in contemporary debates about mental illness, she snaps at him, 'there's no real evidence for the efficacy of anti-depressants', but in one of the last actions of the play, Lorna takes her medication (Prebble 2012, 79). At the same time, Toby's biologistic position that reduces feelings of despair to brain chemistry, what Raymond Tallis has called 'neuromania' (2014), and uncritical belief in the potential of science and industry to cure depression, is not presented sympathetically. Rather, the dystopian scenario suggested here is the reductive pessimism of both positions: the sense of invidious alternatives. Biological determinism, in the form of Toby's biomedical model, does not offer a way out of the impasse any more than the medical or romantic humanisms articulated by Lorna and Tristan. It is not original as such to delineate this predicament, although Lucy Prebble does this elegantly. It is interesting in this production though how Prebble's careful intellectual positioning meets the bold affective technicolour of the Buether and Goold Headlong style and particularly the way that, as with *Earthquakes in London*, non-discursive elements of the staging generate the mixed up feelings and sense of vertigo.

Histories of the condition of vertigo bring together motion sickness, fear of heights and dizziness (Huppert and Brandt 2018). It is somehow more immediate and of the body than the creeping sense of dread articulated by the dystopian form. This is a central metaphor in *Earthquakes in London*. Freya is dizzied by the speed of life all around her and overwhelmed by the prospect of giving birth to a child with the imminent prospect of ecological devastation. She staggers through London buffeted by social, cultural, economic and natural forces. At the end she is left wobbling on the edge of Waterloo Bridge. In *The Effect* Lorna is trapped in bed in a loss of self but spinning in place. In a speech brought on by Tristan's collapse while under her care, Lorna echoes a lecture about the brain given earlier in the play by Toby.[15] Whereas his speech was a witty conference lecture about why he became a clinical neuroscientist, Lorna's starts in the same vein but descends in clipped sentences into a dizzying soliloquy of self-loathing. As she spots each of the neurons firing, in the popular scientific metaphor, she is overwhelmed by dysphoria. 'Why can't you stop? Where are you? Where are you?' she pleads as she pulls the brain apart and prefigures her own collapse (Prebble 2012, 89).

The sense of life speeding up is widely remarked on within Western societies, while containing a number of sociological paradoxes and political ambiguities. The faster technologies appear promising to save time, the more time seems to speed up. Both left and right politics contain progressive and conservative ideological elements as well as arguments for speeding up or slowing down.[16] The immobility within dystopian drama observed by Tönnies in Headlong's work seems to be articulated with a frenzied high pace. Headlong's speculative theatricality across both these and other productions is a deliberate rejection of what Goold has called 'feel-good' drama (Alberge 2012). At the same time, it reflects the paradoxical experience of feeling bad in the midst of apparently hedonic excess, relative prosperity (for their mostly middle-class audiences), and within a popular culture that is often euphoric. The productions are exuberant, colourful and, although not site-specific, have an immersive quality. This quality is partly due to Buether's designs which place the audience within the set in both these productions. It is also significantly a consequence of the music, which in different ways goes a long way to generating the atmosphere of both pieces discussed here.

In *Earthquakes in London*, as I have already remarked, the original production relied on a series of popular, if slightly alternative, anthemic pop-rock songs which generate the dizzy, spinning feeling of vertigo and function theatrically almost like the numbers of the book musicals discussed earlier in the chapter. In *The Effect* the somewhat stranger, more haunting or eerie, feelings were created by Sarah Angliss's original electro-acoustic composition. Surprisingly, most reviewers fail to remark on this feature of the production, perhaps because they are falling over themselves to read the play as scientific debate. Quentin Letts (2012) writing in the *Daily Mail* derided it as an 'ambient hum, of the sort that wants you to think "hey man, high-tech"'. This is rather rudely dismissive of a complex soundscape that runs throughout the piece. However, it does reflect the sense that certain of its sonic features pervade and reference a cultural experience associated with science and technology to the point that they can be read as cliché.

Angliss intends the music to generate a sense of discomfort or the uncanny. This is partly created through found sounds associated with biomedical technologies. The processed samples include: 'scraped metal, struck glass vessels, shaken pill boxes... sonified EEG data... and processed field recordings from the inside of an MRI scanner' (Angliss n.d.). Angliss processes these sounds into a score that is between musical composition and sound effects using software (Max/MSP patches) that combine random generation of effects with compositional functions such as repitching and repetition. *Playing Truant*, for instance, features bubbling electronic notes that seem to repeat but according to a pattern that is hard to make out. Over the top of the bubbling notes is a soaring whine that cuts through or breaks out beyond melody or rhythm. Angliss has referred to it as 'a certain coldness' (2018, 322). The score has coherence because of the way Angliss's approach developed fragments of sound that had a 'family resemblance' to each another. The family resemblance reaches out to a wider cultural soundscape familiar within Hollywood and arthouse science fiction films such as Cliff Martinez's score for the 2002 remake of *Solaris* or, further back, the original 1972 score by Artemyev or Louis and Bebe Barron's work on *Forbidden Planet* (1956). Angliss's approach responds to both the practical context of making theatre where music needs to be rapidly modified and adapted in rehearsal, and the context of the clinical setting through the found sounds. Her algorithmic composition, not strictly a score, constitutes the attempt to remove or distance the piece from instrumental music and the obvious interventions of a human musician (Angliss 2018).

More extraordinarily at the National Theatre production Angliss went further in weaving into the audio 'a deep, pure infrasonic note' (Angliss n.d.). Infrasound, at below 20 Hz, is below the frequency that humans can hear. Angliss refers to it as between music and sound effect: 'extreme bass sound that's on the cusp of perception'. Some extremely large organs are built with a pipe that produces (inaudible) infrasound supposed to generate a feeling of 'awe' (Holden and Bhattacharjee 2003). Infrasound has often been associated with or even offered as an explanation for the experience of the ghostly. In the musicals discussed above, the music offers an alternative or counter-point logic or rationality to that of the narrative. Here it goes further in attempting to delineate the edges of understanding and perception.

Angliss's sense of something missing from the music, the sense of its coldness, is an embodiment of the debate at the centre of dialogues between the characters in the play. Both pairs of characters wrangle over the place of mystery in their worldviews. Toby accuses Lorna of trying to hang on to it, equating mystery with mystification. Connie professes to be at ease with herself as purely physical: 'we are our bodies, our bodies are us… there's not something more'. Tristan wonders at Connie's capacity to live without mystery and regards her scientific materialism as 'a cold way of looking at a person' (Prebble 2012, 34). On one level, the dialogue coherently articulates different philosophical positions on consciousness. However, the theatricality of the speculation also registers the 'cold', eerie feeling we get when we cannot find a comfortable or comforting way through these positions. Coming to terms with this discomfort, with the edges of understanding, seems at least as important to me as having the dialogue, particularly because, as the reviewers seem to have, it is so often dismissed or taken for granted.

Why is scientific materialism felt as cold? In so far as life is felt as warm, then perhaps it is felt as erasing life, or finitude, as such. Humanism responds to this by asserting the particularity of the human experience. Tristan's sunny (warm) humanism is propped up by some kind of loosely referred to religious belief. To a certain extent, the apocalyptic dimension of *Earthquakes in London* also offers a similar promise. An apocalyptic scenario is both crisis and revelation. The apocalyptic form is established by an animated film prologue to Act 5 which moves the action to 2525 post apocalypse and looks back to the time of the first four acts in a quasi-biblical tone. The narrator tells of a 'new enlightenment' brought to the world by a girl called Solomon (Bartlett 2010, 138). Act 5 has a double temporality in which the action proceeds in the (present) time of the rest of the play, interleaved with scenes featuring Freya in the future. Solomon (or Sullivan) turns out to be Emily, Freya's daughter, who at the end of the play plans to walk biblically barefoot from the countryside to give a speech on Waterloo Bridge, the very spot where Freya fell. It is not clear though whether this happens after the play or indeed only in Freya's dying imagination. There are really two ends to the story. There is a cold material one in which Freya just dies: it is the last action of the play. However, in the parallel future time, we are offered a more speculative alternative in which we see her woken up from some cryogenically preserved state in 2525 by her mother Grace. Grace explains where and when she is:

> When you fell in the river, Freya, you hit your head... when that happens, people become unable to see a distinction between their own particles and those around them... They can understand instinctively that we are all just different recycled pieces of a larger, older creature. (Bartlett 2010, 153–4)

Vital materialists put a sunny spin on the discomfort of materialism and recognition of the Anthropocene by according life to everything in a secular re-enchantment of the planet (Bennett 2001). In *Earthquakes in London*, this appears to be both offered by the speculative ending but ambiguously also withdrawn in Freya's more prosaic death. In *The Effect*, Connie follows her physicalist recognition of the human as just body with an aesthetic defence of the position: 'That's enough. It's like, the world is incredible and beautiful, even though we know there's no god behind it. It's even *more* amazing for that' (Prebble 2012, 34). This response reconciles the acceptance of a scientific materialist view of the universe with feelings of awe, amazement and wonder. It is perhaps a kind of vital materialist sublime that brings together the possibility of aesthetic experience that does not seem to rely on belief, superstition or mysticism. It is, in Rayner's terms, perhaps a 'clumsy solution' that brings art and science together as ways of knowing. However, what this response fails to register is the profound sense of discomfort caused by 'wicked problems' such as anthropogenic climate change or consciousness. In the theatre, it is not necessary to decide between the effects of drugs, the effects of love, sound effects and ghosts. For a while it is possible to suspend choices between invidious alternatives.

Uncomfortable knowledge perhaps cannot be articulated in (even vital) materialist terms alone. As a number of scholars from different perspectives have pointed out (Fisher 2016; Tallis 2014), materialism often comes up as deficient in accounting for experience. The ambiguity of *Earthquakes in London*'s ending testifies to the insecurity of any revelatory resolution. By the end of *The Effect*, none of the characters is properly sustained by materialist accounts either. In Mark Fisher's work on the weird and the eerie in popular culture, he puts it like this:

> The condition of materialists... is of knowing that all subjectivity is reducible to matter, that no subjectivity can survive the death of the body, but of nevertheless being unable to experience oneself *as* mere matter. Once the body is recognised as the substrate-precondition of experience, then one is immediately compelled to accept this phenomenological dualism, precisely because experience and its substrate can be separated. There are ghosts in the machine, and we are they, and they are we. (Fisher 2016, 109)

The feeling of the ghost in the machine, experienced by Dorothy from *The Wizard of Oz* in the twentieth century and by most of the characters discussed in this chapter, is the feeling of science in performance. It is eerie. It is uncomfortable. It is a feeling of vertigo. It is the feeling of incompleteness of the materialist account that often emerges when science begins to tackle new problems such as consciousness. In articulating this feeling, speculative theatricality holds together materialist accounts of experience with their immaterial counterpoints: magic, mysticism, mystery, metaphysics.

For Mark Fisher (2016, 83), the eerie is frequently manifested in films and novels as a sense of the human as alien. By the end of *The Effect*, Lorna and Tristan experience their bodies as strange and alien. As a result of his amnesia, Tristan is a bundle of affects that he cannot interpret: 'My balls ache… I'm scared… I'm hot… I feel sick… It hurts' (Prebble 2012, 93–4). As the play goes on, it becomes harder to make sense of any of the feelings from earlier in the story either. Tristan and Connie fall for each other in the first half of the play in a heady rush of flirtation, tap-dance and gymnastic tricks that rapidly become arousal and sex. The second half destabilises the characters' and potentially the audience's understanding of what the apparently familiar theatrical love-making actually means. By the end even the most committed materialists on stage are having their doubts even if they are sublimated onto reflections about the environment: 'Toby: It's crazy weather today. Can't decide anything' (Prebble 2012, 97).

Speculative theatricality constitutes feelings such as vertigo or the eerie within specific sets of collectively observed conventions. The use of infrasound by Sarah Angliss, who has an extensive track record of experimental creative work between art, science and technology, is a particularly interesting use of a little understood theatrical technology to register the limitations of understanding. It is appealing as it seems to push beyond more established conventions such as dystopian or apocalyptic forms or indeed the doubled trajectory of the book musical. Perhaps this is its limitation too though, in that it is not theatrical enough. In its innovative form it becomes rather too hard to read at all as theatrical and just melts into our experience of the 'ambient hum' of the world, as Letts put it. What can perhaps go missing is what Mark Fisher elsewhere (2013b) identified as the 'crackle' in recorded music. We need to be able to locate strange feelings of uncomfortable knowledge within an experience of temporal disjuncture. Theatricality provides a range of resources for locating strange feelings in its situated and dynamic conventions. These may rest in the possibilities for time-travel through dramaturgy or the use of different media, by conventions that articulate contradictory or ambivalent affects and for contrasting feelings on stage with feelings in the audience. In part, these possibilities may emerge from the apparent potential for sensory contradictions and ambiguities implicit in an audience constituted by their capacity to hear, while sitting in the theatre, a place for seeing. Such sensory and epistemological confusion, perhaps compounded by the notion of speculative theatricality, is at the heart of its potential for knowledge repair. Science in performance offers the possibility to suspend making choices between invidious alternatives in response to uncomfortable knowledge through speculating from a position that is out of time and out of place.

Notes

1 Pignarre and Stengers are challenging the equation of rationality as such with reason under particular hegemonic conditions. This resonates with Mark Fisher's critique of 'Capitalist Realism' that articulates the sense in which there appears to be 'no alternative' (2009). Their argument could also be usefully articulated with feminist critiques of hegemonic reason within political communication, such as that by Iris Young (2002), Laclau's (2018) notion of 'populist reason' and Mbembe's (2017) 'black reason'.

2 Steve Bottoms (2012) discusses the ways that a series of British plays engaged with climate change 2009–11, particularly in their representation of climate scientists generally but also examining how they drew inspiration from particular scientists and political narratives. He reflects on how they somewhat expanded the cultural forms through which debates about climate science were being conducted. My argument in this chapter is intended to build on this work and indeed on that of Lavery and Finburgh who in their edited collection (2015) make an argument for the consideration of a much wider array of plays, notably 1950s absurdist drama, in relation to ecological politics.

3 As my colleague Kate Dorney wittily pointed out to me, it is a scientific musical that shows its methods thereby contributing to the public understanding of musicals.

4 Antecedents to Baum's work might be seen in the German fairy tale *Snow White* via Grimm's 1812 version or Lewis Carroll's 1865 novel *Alice's Adventures in Wonderland.*

5 Elphaba seems to be a playfully loose abbreviation of L. Frank Baum the author of *The Wizard of Oz.*

6 The scientist as over-reaching is often discussed as a central feature of science on stage for instance by Roslynn Haynes (2003) and Kirsten Shepherd-Barr (2006).

7 A paradigmatic example of the former may be *Frankenstein* in novel or theatrical form. Brecht's *The Life of Galileo* is often read in the (Marxist) political terms of the latter account although can be staged as a moral dilemma for the scientist.

8 The idea in the *Wizard of Oz* and *Wicked* that witches can be 'melted' by water seems to obliquely refer to the water tortures used in historical witch-hunts.

9 The story of the persecution of the Animals, particularly Doctor Dillamond, in *Wicked* can be read as an allegory of the persecution of Jewish scientists and so-called 'Jewish Physics' in Nazi Germany. The treatment of the Animals also brings to mind debates about biomedical ethics and the condemnation of eugenic research by scientists in Nazi Germany and elsewhere.

10 This was clearly articulating a *zeitgeist* as, despite Milton's poem not having been staged since 1667, two productions (including Goold's and one directed by David Farr at the Bristol Old Vic) opened on the same day in 2004. Goold revived his Northampton production for his first production when taking over the Oxford Stage Company and renaming it Headlong. Phillip Pullman's *His Dark Materials* trilogy was staged by the National Theatre in 2003–4 and also makes extensive allusion to Milton's work.

11 Headlong's production of *The Nether* was a UK premiere of a US play first performed in Los Angeles in 2013.

12 The designer Miriam Buether developed a track record in dystopian visions including a set that disappeared and rotated through 90 degrees, one incorporating a travellator and another where the actors were imprisoned in a glass box (Jays 2017).

13 Here you may detect a nod to *King Lear* that reflects Headlong's envisioning of the contemporary moment through classic narratives.

14 The Cottesloe was refurbished and renamed as the Dorfman theatre in 2014.

15 Delivered with a brain produced from a bucket, both speeches seem to echo Hamlet's soliloquy to/with Yorick.

16 See, for instance, Judy Wajcman's work on the time-pressure paradox (2014) and Benjamin Noys' critique of political accelerationism (2014).

4

Theatre, education and the politics of life itself: Theatre of Debate

Over the first two decades of the twenty-first century, one UK-based theatre company systematically engaged with emergent themes in the biological and related sciences probably more than any other, at least in terms of the number of productions, collaborations with scientists and other experts, and total audiences reached. Unlike the practices discussed so far in this book though, most of Y Touring Theatre Company's work was not presented within theatres. Over this time Y Touring worked largely, although not exclusively, within state-funded secondary schools, most often targeting their work at audiences of secondary school students (11–18). Over a 25-year period from 1989 to 2014, the company toured one or two new productions every year, often of a newly commissioned play, with each play responding to a range of complex themes and reaching several thousand young people and their teachers on each tour. Y Touring was closed as a division of Central YMCA in 2014, but core members of the company have continued to work as the independent charity Theatre of Debate. I devote this chapter to a study of Y Touring's work partly because of the length of time over which the company worked in the field and the consequent depth and breadth of their practical exploration of science in performance. However, my focus on Y Touring recognises the significance of their work not just in quantitative terms but also as a carefully designed dramaturgical approach that they have called Theatre of Debate. The implications of this approach go beyond producing performance. Theatre of Debate also constitutes a distinct form of pedagogy and political practice, where science in performance connects ways of knowing within the context of public education.

Theatre of Debate, as a dramaturgical approach developed, evolved and used consistently by the company, extends from the point of conception of a project through the development of themes, the making of a play and associated educational programme, and its production, normally as live theatre but sometimes also in television drama or filmed performances. The company's practice has also involved a particular approach to the touring and performance of the plays in schools or elsewhere and engagement with audiences both through the live performance and beyond.[1] This dramaturgical approach, I suggest, incorporates: a serious engagement with a range of expertise within a process of co-production; the framing of scientific objects as issues through dramatic narratives and debate; a commitment to the exploration of a politics of life; and the development of particular forms of speculative theatrical debate as pedagogy. As such Theatre of Debate makes a significant contribution to the practice of knowledge repair that I attempt to locate within science in performance. That it has mostly taken place within state education and involved large, diverse groups of young people also demonstrates the democratic possibilities of educational theatre as a way of knowing.

In this chapter, I examine the key features of Theatre of Debate in several ways. In the first section I track Y Touring's 25-year history, giving an overview of their overarching political concerns as reflected in the programmes developed over this period. During this time the company has consistently responded to scientific and technological developments particularly within the life sciences. I argue that this body of work constitutes a creative engagement with what Sarah Franklin (2000) and Nikolas Rose (2007) call 'the politics of life itself'. Theatre of Debate involves processes through which forms of life and the technologies that might be adapting or giving rise to new life forms are understood as sources of power and subject to political contest. I go on to discuss Theatre of Debate as an ethopolitical dramaturgy bringing together the politics and ethics of the life sciences through dramatic narratives, examining the way the practice has engaged with genetics and genomics. In the second half of the chapter, I extend my discussion of Theatre of Debate as a dramaturgical approach, outlining further aspects of the process and how it constitutes an intervention into education that values and promotes embodied knowledge, what I call, following Rose, 'somatic expertise'. In these sections, I describe the development of a specific project dealing with the politics of neurotechnologies, called *Stunted Trees and Broken Bridges*, and discuss a range of responses to the work.

Theatre of Debate

There are two driving concerns that connected the work of Y Touring over 25 years. These concerns constitute a productive tension central to the dramaturgy and politics of the practice. On the one hand, the practice shows a sustained commitment to engaging with young people, from 9–18 but primarily within the 14–18 age range, manifested both in the concentration on touring to secondary schools and in representing young people and their concerns through dramatic narrative. This engagement with young people has been extended through debates, discussions, workshops, programmes run at the venue One KX,[2] collaborations with youth theatres and media activity. All these activities have exemplified a particular concern with supporting young people's sense of their own position within a pluralistic, changing society. On the other hand, Y Touring's Theatre of Debate has also sustained a sense of curiosity for emergent objects from science and technology. Many of the plays produced or developed by the company introduced or even proceeded from specific concepts, diagnoses or technologies often associated with biomedicine. In the 1990s this included HIV/AIDS, Friedreich ataxia, pre-implantation genetic diagnosis, xenotransplantation, depression and genetic modification. In the 2000s this meant stem cells, organ transplants, behavioural genetics, post-traumatic stress, dementia, animals in health research, climate change, surveillance, clinical trials and electronic patient records. In the 2010s they tackled pharmacogenetics, cancer, neurotechnologies, nutrition and diabetes. While live theatre remained at the core of the company's aesthetic, they experimented with a wide range of media including electronic voting technologies, various web platforms, social media, broadcast television, video/DVD production, cinematic screenings, virtual reality and others, often finding ways of integrating these technologies within the live programmes.

Theatre of Debate owes much to the Theatre-in-Education (TIE) movement that in most histories dates from the 1960s.[3] Nigel Townsend, who founded Y Touring in 1989 and remained artistic director over the whole 25 years, had trained at Bretton Hall on the pioneering Drama Education course run by John Hodgson, going on to develop his skills as artist and educator as an actor-teacher with Coventry's now famous

Belgrade TIE in the early 1970s and working for many of the other leading TIE companies. As has been well documented elsewhere, Belgrade TIE pioneered an approach and set of principles that became a kind of aesthetic and political benchmark for subsequent practices in educational theatre and applied theatres more broadly (Jackson 2007; Nicholson 2009). Townsend was a Belgrade company member on programmes such as *Pow Wow* (1973), an interactive narrative for 7- to 8-year-olds set in a Wild West show addressing racism and colonialism from the points of view of settlers and indigenous peoples, and the trilogy of plays about environmental issues *Rare Earth* (1973), that can perhaps be seen as a specific precursor to Y Touring's work with socio-scientific themes.

Townsend left Coventry and moved to London, via theatre jobs in different parts of the UK working for some significant companies including Gay Sweatshop, Unicorn Theatre and Cockpit Theatre. In 1987 he was offered a job developing a new strand of performance practice for the Central London YMCA club. While Y Touring often had a somewhat uneasy relationship with the Central YMCA management, the practice clearly maintained an affinity and a relationship with the youth work of the YMCA. The Central YMCA was the first branch of the international Christian youth organisation which maintained its historic concern with young people's health and fitness. Since its foundation in 1844 the YMCA had developed a worldwide range of youth work practices appealing to 'mind, body and spirit'. The association has been credited with the development of basketball and volleyball in a form of what has sometimes been termed 'muscular Christianity' (Putney 2001). In the early twenty-first century, this historic mission was interpreted by Central YMCA principally via their management of a large fitness gym in central London (Central YMCA 2016). In many ways, Y Touring's attempts to engage young people with the politics of the mind and the politics of the body, albeit within a more secular frame of reference, was very consistent with this ethos, even if the YMCA management might not always have appreciated this. This ethos was combined in Y Touring's practice with some of the philosophy of the TIE movement. At Belgrade and elsewhere, TIE had been underpinned by a combination of progressive, child-centred educational philosophy and sometimes revolutionary Marxist politics. The clear legacy of TIE in Y Touring's work is evident in the concern with the development of particular forms of dialogical theatre. Y Touring's work therefore combined a youth work sensibility – a concern with young people's self-realisation – with the dialogical pedagogy of TIE.

As has already been referred to in this book, the HIV/AIDS pandemic has played a significant role in shaping the politics and aesthetics of science in performance and this is also born out in the early history of Y Touring. The company's first production was their 1989 tour of *The Inner Circle*, a play by US playwright Patricia Loughrey that had initially been developed and produced by Ed Decker at the New Conservatory Theatre Center in San Francisco. The play toured in the USA in the 1980s and subsequently in the 1990s became a popular play for US high school productions (Educational Theatre Association 1992, 1994). It has also been translated into a number of other languages and performed internationally in over 500 productions ('Patricia Loughrey – Plays' n.d.).[4] The play responded to the HIV/AIDS pandemic focussing on the experiences of four young characters. The story contains a number of facts about transmission of the virus but within an engaging narrative recognisable to its intended audiences of young people. The perspectives of the characters also reflect a range of social attitudes typical of the time including the stigmatisation and discrimination around AIDS. Y Touring's UK version of the play adapted it for its British audiences and incorporated a workshop that included hot seating the characters – a technique used by Belgrade and other TIE groups which remained a key feature of Y Touring's practice. The production gained support from Crusaid, an HIV/AIDS charity that subsequently merged with the Terence Higgins Trust, and the Hampshire and Dorset Health Authorities. It toured UK schools for eight years, even reaching the Orkneys, showing the sustained importance of the theme but also how extraordinarily successful the theatre project was in addressing what was a very difficult issue for schools, particularly at that time.

There are key aspects to this first production that are important in understanding the significance of Y Touring's work. Firstly, that while continuing to draw on the artistic and educational practices of the TIE movement, *The Inner Circle* represented a distinct form of educational theatre through the way it engaged with health, both in terms of biomedicine and young people's sense of self. Secondly, Y Touring's approach to theatre and education was a co-production process involving close collaboration with non-arts charities and health authorities. Both these dimensions of the practice reassemble practices of educational theatre in relation to a growing sense that the political rests as much in individual lifestyle choices – matters of life and death – as it does in collective affiliations. This perspective reflects sociological shifts such as the changing role of women, globalisation, civil rights, gay rights and identity politics. Such shifts have been influenced by and expressed through notions of biopolitics. Michel Foucault claimed that, in the modern era of biopolitics, 'the life of the species is wagered on its own political strategies' (1979, 143).

Biopolitics or the politics of life are inextricably entangled with changing conceptions of life emerging from the biosciences and the changing technologies that impinge on and, to a greater or lesser extent, influence the way we live. Sarah Franklin describes such changes as 'transformations and realignments of life, labour and language' (Franklin *et al.* 2000, 193). The YMCA's historical approach to youth work combining architecture, physical exercise regimes and spiritual wellbeing and the TIE movement's approach to embodied learning across the school curriculum both demonstrate reflexivity towards biopolitical experience. Both movements worked within and with established institutions of church and state but with creatively innovated sets of political practices that may not immediately have been recognised as such. Central YMCA evolved from its initial nineteenth-century origins, in the bedroom of a draper George Williams near St Paul's Cathedral, to an organisation open to all faiths and none (Gilderson 2019). Its biopolitical practice has been dominated by a commitment to sporting exercise and physical training, but it has also integrated this with other kinds of artistic and educational activity.[5] Y Touring's practice emerged from this intersection. The company brought to this a consistent engagement with the biosciences presaged by the emerging environmentalism within TIE reflected in *Rare Earth* and the health activism around HIV/AIDS reflected in *The Inner Circle*. As such the politics of the company's practice seemed to track closely the way the 'politics of life itself' evolves as an iteration of biopolitics.

Y Touring's work though cannot simply be seen as health or environmentalist education or activism, even if some of their projects have engaged with these fields. Their intervention into the politics of life, rather than espousing a particular activist agenda, has been to open up emergent objects, ideas and technologies to ethical evaluation through the Theatre of Debate process. The consistent position has been to resist premature closure to such evaluation, whether this might happen through ideological, moral or scientific rationalisation. Theatre of Debate is therefore an 'ethopolitical' practice. Nikolas Rose explains ethopolitics as allowing 'the possibility of opening up the evaluation of forms of life and self-conduct to the difficult and interminable business of debate and contestation' (1999, 192). Theatre of Debate's engagement with the biosciences and concern with emergent objects and forms of life has not come with an easy acceptance of scientific or other discourse that might assume priority or authority over articulating the facts of such life. Staging the 'politics of life itself' even involves resisting premature closure over what might constitute life and an examination of the proposition that life itself is not (or no longer) co-extensive with living beings.[6] Theatre of Debate constitutes a particular pedagogy that cuts across knowledge boundaries, as defined within schools via curriculum or teacher specialisation, and has mobilised diverse forms of expertise in reworking both pedagogy and curriculum.

The rise of genetics and neuroscience has played a central role in the contemporary politics of life and, as such, provides a thread through Y Touring's practice. Theatre of Debate has sustained an exploration of genetics and genomics from the 1995 production and tours of *The Gift* by Nicola Baldwin, dealing with the ethics and politics of genetic screening, through to and beyond the 2011 production of *Dayglo* by Abi Bown that examined the potential of genomics to personalise drug treatments.[7] These two productions show something of the evolution of thinking and applications of genetics over the period, and how these conceptions intersect and are co-produced with popular narratives and particular ethical debates. *The Gift* featured ethical issues that arise from single gene disorders, relatively rare inherited autosomal recessive conditions, which can be passed on from parents to children by a particular genetic mutation. This illustrates the so-called 'gene-for' paradigm whereby genetic factors are seen as determining health or particular traits. If you inherit two copies of the gene, then you will definitely develop the condition. *Dayglo* featured a character whose brother has sickle cell disease, another autosomal recessive single gene condition, but this is juxtaposed with the story of a character with a faulty BRCA1 gene, linked to an increased susceptibility to breast or ovarian cancer.[8] So-called faulty or mutated BRCA1 and BRCA2 genes can result from many different small changes to DNA, called single nucleotide polymorphisms (SNPs), but even inherited forms of breast cancer depend on a number of factors beyond these genes; so testing positive for BRCA 1 or 2 only means an increased probability of developing breast cancer rather than its certain onset. Breast cancer can of course be developed without either of these genetic profiles. Neuroscience and its evolution, alongside changing conceptions of mental health, was also a recurrent theme from the 1996 production of *Cracked* by Nicola Baldwin through to the 2012 play *Stunted Trees and Broken Bridges* by Ben Musgrave.[9]

Genetics and neuroscience are both clearly burgeoning areas of science and technology. They are also co-produced with and by a popular imaginary with its own discourses, narratives and iconography familiar to scientists and non-scientists through popular media, internet memes and entertainment. As José van Dijck (1998) has argued, 'popular images of genetics are not created by either social or professional groups, but evolve from the continuous interaction between these groups in the realm of culture'. These images are not fixed but transform over time through this interaction. Van Dijck goes so far as to say that the evolution of genetics should not be seen as a series of discoveries by scientists but as 'conceptual shifts, scaffolded by images and imagination' (1998, 179). It is the perceived gap within this co-produced imaginary, so between the understandings and feelings of scientists, policy-makers and other sections of populations, that has driven public engagement policies, performance pressures and funding discussed in Chapter 1 of this book. Theatre of Debate has operated within and across this gap, dramatising a process of co-production in a way that connects up scientific conceptions of genetics or neuroscience with a more popular imaginary. In this way, Y Touring's 'genetic imaginary', to use Sarah Franklin's (2000) term, or indeed what might be called their neuro imaginary, are distinct from those circulating both within scientific journals and within mass media. In fact, Theatre of Debate might be better termed a process for imagining genetics or neuroscience rather than a coherent or consistent 'imaginary' because of the way it resists closure or the imposition of a particular frame. It exposes the process as process and gives audiences purchase on it, deliberately inviting their imaginative participation. This is a significant departure from dominant narratives and imaginaries circulating across much of the popular media.

Franklin, for instance, has argued that blockbuster Hollywood movies, like *Jurassic Park*, tend towards a tone of 'moral denunciation' in their treatment of science, even as the film and its merchandise celebrates it and performs its own commodifications of it (2000, 202). She argues that this is facilitated by the particular ways that the film manipulates 'the axis of the visual', inviting the audience to look behind the scenes 'speaking precisely to the knowledge of its audiences that there is more going on than meets the eye' (2000, 216), feeding off and reflecting popular ambivalence towards genetics that I have discussed already in this book. The speculative dimension of the film, the exploration of the possibilities of the technology, is closed off, in ethical if not economic terms, by the moral certainty of the conclusion. Its social critique is then somewhat undermined by the way the film and its spin-offs exploit the profitability of biotechnology. Audiences are positioned as cynical 'public witnesses' of science but, at the same time, as consumers and potential dupes of global biotech. Similarly, much media reporting of neuroscience speculates about the potential of neurotechnologies such as functional magnetic resonance imaging (fMRI) but tends to do this within a kind of closure about the knowledge, rarely addressing the limitations of the research or the technology itself. This tends to reinforce notions of 'neuro-realism' and 'neuro-essentialism', whereby brain imaging data is presented both as a kind of 'ultimate proof' about a phenomenon and as a novel essence of humanity, superseding notions of personhood, the self or the soul (Racine *et al.* 2010, 728). At the same time, such media coverage also tends to reach premature ethical closure about the technology through an exploration of benefits or more rarely its potentially damaging consequences.

Ethopolitical dramaturgy: staging the politics of life

Theatre of Debate provides a contrasting dramaturgy, a distinct way of positioning its audiences and consequently a somewhat different political practice to this otherwise dominant model. It shares a mode of speculative theatricality, examining the (often very near) future possibilities of emergent technologies and changing conceptions of life. However, rather than a process of closing off, Theatre of Debate relies on an iterative process of opening up genetics and an ethopolitical perspective and thereby an opening up of possible futures. This hopeful yet pragmatic perspective, which displays a concerted resistance to economic or technoscientific determinisms, is reflected in Y Touring's sustained commitment to working primarily for young people and within education. Such opening up relies on dramaturgical processes that are consistently reflexive and responsive towards the shifts in contemporary biopolitics. Rose (2007) has examined these shifts, which he refers to as 'mutations', in some detail and there are key elements of his analysis that are useful in exploring Y Touring's Theatre of Debate.

For Rose, shifts in contemporary biopolitics constitute 'styles of thought' that are manifested in technologies and shape emergent forms of life. The first of these trends, that Rose calls 'molecularisation', connects many of the central themes of different Theatre of Debate projects. To some extent, this is a shift in scale, life is visualised and conceived by biomedicine at the molecular level rather than the whole organism or set of organs. If the narrative concerns of educational theatre were hitherto concerned primarily with (whole) human bodies, communities and societies, contemporary biopolitics forces attention towards these bodies as made up of components that might be recombined, exist outside the body and be subject to a variety of medical and other interventions. Molecularisation relies on a whole range of genomic and information technologies and on technologies that can visualise the body at the molecular level. Such technologies and their connected molecular concepts and imaginaries pervade Y Touring's plays. For instance, brain imaging, its limitations as representation and how it sets up a particular imaginary about life is a central concern of *Mind the Gap* and *Stunted Trees and Broken Bridges*. DNA sequencing technologies provide part of the premise for *The Gift*, *Dayglo*, *Making Astronauts* and other plays. A second and linked trend is what Rose calls 'optimisation' (2007, 15). Biomedicine in the twentieth century was primarily concerned with establishing the norms of life, then pathologising and treating departures from these norms where possible. Biomedicine in the twenty-first century identifies susceptibilities of and possible enhancements to the body and so proliferates the potential for technological interventions among populations who might previously have been considered healthy. Theatre of Debate plays tend to pose questions about technologies of optimisation, whether this is the potential for cognitive enhancement dealt with in *Mind the Gap*, the neurotechnologies in *Stunted Trees* and *Broken Bridges*, or the susceptibility to particular inherited conditions dealt with in the genetics plays.

Questions of optimisation and molecularisation are often interlinked in the dramaturgies of these plays. For example, the narrative of *The Gift* by Nicola Baldwin, first produced by Y Touring in 1995 and toured again to UK schools in 2000, relies on a set of technologies, developed in the 1990s, whereby the gene for a particular recessively inherited condition can be identified in adults who have the condition or carry the genetic mutation for it.[10] Pre-implantation genetic diagnosis emerges from this technology in combination with existing infrastructures for assisted reproduction (IVF). It enables would-be parents to test embryos for a condition which may be present within their family history.[11] By using this process and only reimplanting embryos without the gene, such conditions might potentially disappear from families and societies. Families undergoing such processes normally receive genetic counselling that explains the condition, its mode of inheritance and some of the other factors and risks that they might need to consider in deciding whether to use the technology. *The Gift* features these technologies and a speculative projection of them, but also, by playing out the narrative across three generations, it tracks how particular decisions made on a molecular level might have implications for individuals, families and societies.

The Gift tells the story of three generations of the Kay family. It opens in the year 2030 as Mark Kay, nearing his 16th birthday, falls out with his father Ryan. The story then flashes back to the year 2000 when Ryan was himself a teenager.[12] Ryan's sister, Annie, is a promising footballer but starts to experience 'little gaps' and loss of balance (Baldwin 1995, 10). Annie is eventually diagnosed with Friedreich ataxia, a rare inherited degenerative condition of the nervous system. As a doctor later explains to Annie and her mother, 'Friedreich ataxia is passed on in families in a way we call "recessive". A recessive disorder is going to be relatively rare, as you need both parents to be carriers' (1995, 12). That is to say, Annie has inherited two copies of the gene from her parents who are both carriers and therefore has the condition. The subsequent implications of this are presented in the imagined futures of 2014 and 2030. In 2014, Ryan has discovered that both he and his wife Jennifer, with whom he wants to have a child, are carriers. In order to ensure that his child does not inherit the condition, he selects an embryo without the gene before it is fertilised in vitro and then implanted in Mark's mother, Jennifer. Mark is aware of this but in the opening scene he has discovered that his father also selected a genetic profile which accounts for his sporting prowess – the gift of the play's title. Mark has found out because, in the 2030 of the play, 'at schools all over Eurasia' 16-year-olds routinely undergo genetic profiling. He is sent his profile in advance and is furious because, as he says, 'the one thing I ever thought I was any good at turns out to be just another thing you gave me' (Baldwin 1995, 4).

The Gift imagines the body at a molecular level, where the health of a body is determined by a particular sequence in a particular section of a chromosome. Friedreich ataxia results from a repeated triplet of nucleotides, GAA in the familiar four letter script that represents the structure of DNA (Campuzano *et al.* 1996). The characters in the play often talk about a 'rogue gene' or 'wretched gene' that has wrecked the family. The gene has developed a malign agency independent of the people it makes up. Within the narratives, the characters' experiences of their bodies as whole are juxtaposed with the way the body can also be visualised and perceived at the level of organs, cells or molecules separable or independent from the body. Facing Mark's predicament in *The Gift* means joining up molecular understandings of life with a molar sense of self. It also requires an appreciation of his relationships and family history. A genetic imaginary conjured up by the potential of screening technologies sits beside a brother's relationship with his sister, a nephew's memories of his auntie and a fictional adolescent's particular sense of self.

The narrative builds on contemporary possibilities in genetic screening (for conditions like Friedreich ataxia) and more speculative notions such as the possibility to select an embryo for sporting prowess. Optimisation of life is therefore framed in a number of different ways within the play: as life without Friedreich ataxia, life without the prospect of passing on the condition to children, life with particular athletic talents. The play unpacks the way that the idea(l) of optimal life is a function of family, society and technology in different conjunctions at different times. Before Annie is diagnosed, she is a successful footballer and affectionately teased by her brother for this perceived challenge to gender norms. The new technologies are themselves depicted in the play as transforming from a contemporary clinic to a futuristic genetic screening centre where genetic profiles can be conjured onto computer screens. Visions of optimisation also repeatedly generate disagreement. Ryan and his wife Jennifer disagree about genetic selection – she wants a 'natural pregnancy' – and he can only just about persuade her to select out Friedreich ataxia. Unbeknown to her, he also selects for athletic ability, provoking his son's angry outburst years later.

José van Dijck, writing around the same time as *The Gift* was first produced, outlined a 'recurring paradox… between increasingly hybrid technologies and scientific practices, and their monolithic or purified representations' (1998, 194). As technologies become composed of more and more elements – private and public sector research, biological processes, information technologies, social infrastructure, multiple professional specialisations – the imaginary can be dominated by a single vision and a simplistic moral evaluation either approving or condemning. *The Gift* cuts against such purification, neither constituting critique nor promotion. It also locates both technologies and evaluations of them within changing socio-cultural contexts, unfixing them and throwing them open to question. Within the relatively short and uncomplicated narrative of *The Gift*, genetics is situated in a public health service, private corporate offices, a family home and on a tennis court. It has eugenic potential as well as the potential to avoid a cruel and debilitating disease and is intertwined with complex socio-cultural idea(l)s about the optimisation of life and the perfectibility of the human body.

Ethopolitical questions emerge through Mark's and his parents' fictional experience. His parents' choices around genetic screening and decision not to reveal this to Mark are put into question. Ryan's decision, and the audience's understanding of this, is shaped by his relationship with his sister Annie and her experience of Friedreich ataxia, but clearly also by views on other issues such as his apparent desire for a successful and athletic child, perhaps itself also shaped by his memory of his sister's thwarted ambition as a footballer. Ryan's capacity to undergo the genetic screening process, interpret the genetic profiles and intervene in it without his wife's knowledge is a result of his income and training as a geneticist, a career choice clearly influenced by his sister's story. Mark's anger when he finds out what his father had done is clearly driven by the disruption to his sense of self caused by the knowledge. There is no sense in which the choices made by the characters are capricious or lightly arrived at and so these choices cannot be simply evaluated.

The dramatic structure of *The Gift* opens up an ethopolitical perspective on genetic screening partly through the way it spreads the narrative across three generations and across 30 years, enabling the debate to examine implications of technology across a significant timespan, through a play and discussion that mostly takes less than two hours. It also opens up the perspective by playing out the processes, transformations and implications of molecularisation and optimisation on characters that are otherwise ordinary, depicted in everyday relationships with parents or siblings, making everyday choices about having children. Unlike *Jurassic Park*, or many other popular treatments of genetics, the central characters are not exceptional figures, motivated by hubris or living within a dystopian autocratic regime pre-empting ethical evaluation.

If *The Gift*'s approach to opening up an ethopolitical perspective relies on the story playing out over generations and decades, Abi Bown's *Dayglo* makes use of parallel narratives crossing generations and social divides. In the play, two young people, Stella and Noel, meet each other in the fried chicken takeaway shop where Noel works. Stella's mother Evelyn has been diagnosed with breast cancer. Stella has been separated from her mother whose lifestyle as an iconoclastic rock star seems to have affected their relationship. After Evelyn's diagnosis though, Stella comes to see her mother and they both begin to rebuild their relationship. At the same time Stella and Noel become friends and are attracted to each other. They have things in common: they are the same age and share an experience of coping with illness in the family – Noel's brother has sickle cell disease. As with *The Gift*, the ethopolitical perspective emerges in *Dayglo* as the possibilities and limitations of biomedicine open up within the narrative and are faced by the different characters.

Dayglo plays with the perspectives of different generations. Evelyn lives life according to an anarchic ethos defined by her career as a punk singer.[13] She is confrontational in her dealings with the medical establishment and although she has a mastectomy, she is resistant to further testing and treatment. In fact she does not want to know any more about her condition. 'Never Google, never worry', she says (Bown 2012, 10). Her daughter Stella seems very different. Faced with her mother's condition she ploughs through internet forums and social media searching for information about breast cancer. She is keen for her mother to have a MammaPrint – a genomic test that can predict if a cancer is likely to spread – and a BRCA test to find out if the cancer might be inherited. She also researches their family history looking for the illness across previous generations.

Stella and Noel's relationship is tested by the apparently different status accorded to breast cancer and sickle cell disease. Noel is frustrated by what he sees as a lack of availability of screening and treatment for sickle cell, which he links to the marginalisation of those with the condition.

Sickle cell disease is prevalent among black and minority ethnic communities.[14] Noel suspects that research is curtailed by a lack of potential profit to be made from treatments. He speculates on the likelihood of personalised medicine being developed for his brother Josh in a witty dialogue with Stella comparing the economics of pharmacogenetics to the economics of fried chicken. At the chicken shack where Noel works, Hester from the pub likes Peri Peri sauce, but the owner of the takeaway is not going to go to the trouble of making it, especially as Noel points out because: 'Hester's a cleaner – she can't afford to splash out on some fancy chicken recipe, she'll eat southern fried like the rest' (Bown 2012, 16). As in *The Gift*, genetics is not monolithic. Within the scientific discourse, genetics and genomics are differentiated in the tests for different conditions. The technologies are constituted by science *and* economics. They save or prolong life as well as make profits for pharmaceutical corporations.

In *Dayglo*, *The Gift* and indeed the other Theatre of Debate plays, ethical perspectives are interwoven with explorations of hybrid technologies, economics, ethnicity, age, family background, personality, even affiliation to popular sub-cultures and other aspects of identity. Knowledge developed by the characters in these plays draws on professional, scientific expertise but equally from the experience of patients, carers, families and friends. Knowing is resourced by a symbolic and metaphorical vocabulary drawn from science but also from sport, fast food and music that all shape the feelings of the individual characters and popular feeling more broadly. Political agency in the plays emerges from the ability to mobilise this vocabulary in relationships between non-experts and experts and between non-experts as well as negotiating and, to some extent, unfixing the symbolic vocabularies of authorised scientific discourse. That scientific discourse relies on metaphor and symbolism at all can be rather opaque. Jackie Stacey has drawn attention to the way that biomedicine tends to use metaphor covertly, 'effectively covering its figurative tracks' (1997, 52). The proliferation of metaphor, analogy and simile in the plays helps to uncover these tracks. In the plays, the vocabularies are very often in the hands of non-professionals. In *The Gift* Annie explains the inheritance patterns of Friedreich ataxia to her brother using coins. In *Dayglo* Noel has his fried chicken analogy for the bioeconomics of pharmacogenetics.

Political activism within both *The Gift* and *Dayglo* takes novel symbolic and associational forms – a key theme of the next chapter in this book. In *The Gift*, Annie joins a group of patients with Friedreich ataxia, who replace the friendship and collective affiliation of her football team. Annie, Ryan and their mother campaign for rights to genetic testing for under 16s on behalf of Ryan and other young people. Ryan and Jennifer fall on opposite sides of debates about genetic testing but still have a child together. In *Dayglo*, Stella goes online to gather information about breast cancer and makes friends with people through social media whom she connects with over her relationship to breast cancer and genetics. Noel does a sponsored run for sickle cell research. Ethopolitical affiliation and ethopolitical action are depicted as taken in relation to and through the body.

Perhaps the key respect in which Theatre of Debate functions as pedagogy and politics is the way that it brings together different conceptions of the body. It connects the scientific or biomedical body – subject to the institutional power of schools, hospitals, research centres and private industry – with the felt body as experienced in emotional or affective terms and the social body, constituted in relationships between family or friends. Ethopolitics demands a merger or intersection of these frames. Such a political move is of course in line with broad trends in feminist theory and practice which have attempted to challenge established distinctions between public and private spheres (Fraser 1990). It is also consistent with the politics of feminist sociologists who have challenged how societies value and frame feeling as labour or work (Hochschild 1979). The particular contribution that Theatre of Debate makes as educational practice, however, is in the way that feeling is constituted and valued as a way of knowing. Knowledge derived from feeling interacts with and is held together with other ways of knowing through the dramaturgy of the plays and indeed through other features of the practice I explain in the next section.

Many of the characters in Y Touring's plays exemplify ways that felt knowledge and disciplinary knowledge of biomedicine are synthesised in ethopolitics. Rose calls such emergent knowledge 'somatic expertise', registering the way it is knowledge held in and of the body but also increasingly valorised as expertise that can be gained and sold as labour (2007, 28–9). Somatic expertise, for Rose, gives rise to what he calls 'pastoral power' – power constituted in 'the relation between the affects and ethics of the guider… and the affects and ethics of the guided' (Rose 2007, 74). Such relations are frequently staged in the plays, for example in the kind of formal genetic counselling between the Kays and clinicians in *The Gift* or indeed in the informal discussions between characters in *Dayglo*. This expertise clearly includes scientific knowledge: the understanding of principles of genetic inheritance and some of the nuances of the genetics of particular conditions. However, it might also include an appreciation for how ethical positions in relation to particular scenarios might be informed by particular emotional dynamics, family experiences, national histories (e.g. of eugenics), religious beliefs and other factors, as I have tried to outline in relation to *The Gift* and *Dayglo* in particular. Brian, an oncology nurse, treats and befriends Evelyn in *Dayglo* and his apparent wisdom and skill are rooted as much in his campy humour and ability to relate to both Evelyn and Stella as they are in his technical understanding of cancer genetics. In an opening scene of the play, before Stella returns to Evelyn, he does not so much explain Evelyn's diagnosis in biomedical terms as in relational terms, reflecting on the difficulty of sharing the diagnosis with a 'teen, probably nursing a grudge – the worst kind' and how 'this illness tears families apart'.

The characters represent clinical professionals such as nurses and doctors who would be recognisable to young audiences and whose authority is both extended and challenged within life politics and by their development of somatic expertise. However, they also include the novel authorities who emerge within life politics, such as genetics counsellors and forensic neuroscientists. Importantly, the plays also show the acquisition of such knowledge and power within and without professional roles and institutionally defined frames. In *Dayglo*, Noel has an understanding of genetics gleaned from his brother's sickle cell disease. Stella learns about cancer genetics from social media. Disciplines and institutions have an uneasy hold on somatic expertise and therefore pastoral power. If genetics within Theatre of Debate is not monolithic, then knowledge about it cannot be easily contained within monolithic institutions.

The development of an institutional infrastructure for ethopolitics and somatic expertise in the UK coincides with the history of Y Touring, and indeed the period covered by this book, and has been the subject of much political debate in this time. For example, the UK Human Genetics Commission, a non-departmental public body, was created in 1999 and then abolished by the government in 2010. The independent Nuffield Council on Bioethics, which has been involved in several of Y Touring's projects, was established in 1991 and is sustained by a mixture of charitable and public money. Bioethics as a broader discipline had roots in the much older system of medical ethics. However, medical ethics was primarily a matter for the medical profession. Bioethics has only emerged since the 1970s as a much more inclusive area of debate, where the views of a much wider section of society are considered of relevance, and therefore has much wider implications for education (Chadwick and Wilson 2018).

Y Touring's work has offered numerous stagings of encounters in which pastoral power is produced within the theatrical frame through the exercise of somatic expertise. In doing this, the practice offers insights into how this power is constituted that are hard to access elsewhere, outside personal experience, given that these encounters are almost by necessity held in private. Even professional bioethics according to some commentators has been rather slow to examine what people actually do and think when faced with complex medical scenarios (Borry *et al.* 2005). Such knowledge therefore has unsurprisingly made relatively little inroads into general educational curricula for young people. From the late 1990s there were initiatives to steer the English school science curriculum, particular in the 14–16 age range, more towards an approach that brought in some of what I am calling ethopolitics. Some of the pedagogical challenges identified in relation to this move had to do with how to make philosophical and sociological questions about science accessible to this age group (Millar and Osborne 2000).[15] Theatre of Debate goes some way to meeting this challenge: its significance as pedagogy rests in its mode of producing and framing somatic expertise for and with young people. In the rest of this chapter I will explain how this approach extends beyond the dramaturgy of the plays throughout a creative practice of (etho)political education.

Ethopolitical education: overflowing frames, emerging issues

Y Touring as a company sustained their practice over 25 years because of a capacity to engage with scientists and scientific discourse and develop relationships with the emerging institutions and sponsors of bioethics, most notably in their relationships with organisations like the Nuffield Council on Bioethics and the Wellcome Trust. At the same time as maintaining these institutional relationships, the company accommodated their practice to the shifting demands of state schools in the UK, satisfied the evolving biopolitics of Central YMCA, supported and developed a group of writers and other artists, while also valuing the experience and perspectives of young people and a variety of somatic experts. This may sometimes have seemed like political and epistemological contortionism to the managers of the company. However, I would argue that this process of co-production constitutes Theatre of Debate as an ethopolitical educational practice.

The process relies on the co-production of knowledge within a series of overlapping frames and consistently allows, even prompts, this knowledge to overflow these frames. As I will go on to discuss, with reference to the sociology of Michel Callon (1998), technoscientific knowledge tends to overflow frames. This overflowing gives rise to issues. As subsequent work in the politics of science and technology studies has argued, issues assemble and mobilise publics (Marres 2012). It is the exploration of issues that provides the terrain for politics and debate. Theatre of Debate is therefore not so much an issue-based practice as a practice whereby issues are created. The practice relies on engaging with the multiple individuals and institutions I have just characterised, as participants and audiences, all of whom bring with them their own ways of framing knowledge. Again, holding together this collective is a key dimension of Theatre of Debate as ethopolitical practice. At this point in the chapter, I should acknowledge this reflection emerges in part from my own changing relationship with the company, which itself has been informed by different but overlapping frames for viewing the practice: as funder, as commissioned evaluator and as academic researcher.[16]

I will illustrate and discuss the Theatre of Debate process here with reference to a project conducted by Y Touring in 2012–13, initially entitled 'Brainwaves' and leading to the play *Stunted Trees and Broken Bridges* by Ben Musgrave. As the second to last play before Y Touring closed as a division of Central YMCA and re-emerged as the independent Theatre of Debate, it shows the latest iteration of a process which had become tried and tested by that point. The exploration of a project with a neuroscientific theme here also complements the focus on genetics and genomics from the first half of this chapter. The starting point of most Theatre of Debate projects is framed by an advisory group or in funding applications within a loosely scientific discourse about an emergent object of science and technology – a novel technology, diagnosis or form of life. The Brainwaves project was described at its outset as being about 'novel neurotechnologies'. Such a starting point may, of course, risk a framing according to the discourses and ideology of experts on an advisory group or funding body that comes to determine the practice.

Roger Wooster has criticised the emergence of what he terms theatre in health education in the 1990s for what he perceived as a narrow approach driven partly by project funding instead of ongoing support from Local Education Authorities. For him this led to a tendency to 'teach "about" specific issues determined by the purse-holder' (Wooster 2016, 189). Wooster specifically theorises the pedagogy of 'classic TIE', or what he calls 'conceptual TIE', as developing critical consciousness in terms of the Brazilian educator Paulo Freire's *conscientizaçao*. He equates work that is issue-based with work that is message-driven or framed in a way that is necessarily uncritical and therefore de-politicised. For him, such practice loses its counter-hegemonic function and becomes more closely associated with the didactic banking model of which Freire is so critical. This critique may well be justified in response to some practice, although to apply it broadly to theatre in health education risks underestimating the political dimensions of health and the ways that this theatre practice engages with the politics of life itself. Such a critique also conceives of the 'issue' as necessarily fixed or closed and somehow outside the realm of the political instead of the necessary terrain of political debate.

The Theatre of Debate process does not start from an issue so much as an object, in this case novel neurotechnologies, and the initial framing is not allowed by the process to dominate the practice, in fact the initial project title 'Brainwaves' already offers a different framing. Issues in any case are not outside ethopolitics but constitutive of them. Within Theatre of Debate, issues emerge and evolve through the course of a project that incorporates a number of stages including: advisory group discussions, pre-script workshop, writing, feedback and revision of drafts, rehearsal, performance and pre- and post-performance workshops and debates. Each of these stages makes use of a range of events and aesthetic forms that articulate the issue through: presentations, narrative, characterisation, visual and linguistic metaphor, sound, music, scenography, debate and other modes of interactivity including electronic voting and social media.

In Theatre of Debate, the writer has played a central role in framing the issue through narrative and character development, as I have discussed in the first half of this chapter. However, the writer's script follows the identification of an overall theme - scientific object – by the company's advisory group. A brief for the writer is then normally set up by a workshop process in which a number of experts as well as teachers and young people participate. In some cases, a number of writers were invited to these workshops and pitched synopses to the company. In other cases, the workshop was oriented towards a writer who had already been selected for the project. This was then explored via a workshop which involved ice-breaker activities, a number of presentations from experts followed by discursive and imaginative exercises involving the experts, teachers, artists and young people examining the implications of the presentations and developing possible scenarios or narratives that might emerge from them. This whole process is managed and facilitated by core company members and normally led by the artistic director, Nigel Townsend.

The company has often captured the workshop process for later use by the writer and others via different modes of graphic note-taking. The image in Figure 4.1 by a graphic artist is from the notes of the workshop held in 2012, which led to *Stunted Trees and Broken Bridges*. This image gives an artist's impression – or framing – of the objects emerging from presentations given as part of the workshop. They include

Figure 4.1 Brainwaves graphic note

a variety of potential technical devices including drugs, electrodes and scanners. The notes also gesture towards possible ways that these devices could be framed or assembled as technologies – military, therapeutic, judicial – from which issues might emerge. In the middle part of the graphic there are the images of the electrode and a lie detector test and the quotation 'thought crime is not a crime' that evoke possible imagined scenarios that might include that of *Stunted Trees and Broken Bridges*.

The curating of the presenters, the assembly of the other workshop participants and the facilitation of the workshop are all important dimensions of the ethopolitical dramaturgy. The tendency of these workshops is to multiply the way that the initial object 'novel neurotechnologies' might be framed, and to open up the possibilities for issues to emerge from objects. Expert speakers introduced quite a wide range of different objects into the process and used quite a range of rhetorical strategies, grounding their explanation of technical devices within narratives and posing ethical questions and unresolved dilemmas to the other participants. This rather belies any sense that the role of experts in such a collaborative process is to present value-free facts that artists then make into an emotionally engaging narrative set into a credible, social context. The workshop process is therefore not just to generate issues from a set of objects existing only within the context of the laboratory, but rather to assemble and map objects and emergent issues collaboratively. The expert speakers can be seen as Latourian 'spokespersons' for nonhuman objects – in this case cochlear implants, neurons, electrodes, scanners, etc. – as well as intermediaries for other individuals and groups whose perspectives they might represent, such as individual patients, advocacy groups and other scientists (Latour 2004b).

Figure 4.2 illustrates the experimental, improvised stories generated by and generative of discussions of issues by the mixed groups involved in

Figure 4.2 Brainwaves graphic note

the Brainwaves project, including experts, teachers, students and artists, at the end of the day after a number of presentations and creative exercises. The image gives an insight into the issue framing process prior to the work of the writer but after the presentations illustrated in Figure 4.1. At this point, it is possible to detect objects and issues from the previous image – a scanner, drugs, a syringe – but there are new objects, characters and settings. This image also suggests a reframing of the issues, the still hypothetical statement 'thought crime is not a crime' has been located within the context of post-natal depression or within the life experience of a young beatboxer. The writer Ben Musgrave then worked with this material and his own research in producing a script which was further developed with dramaturgical input from a number of those who took part in the workshop as well as Townsend. Musgrave's script for *Stunted Trees and Broken Bridges* shows many traces of the research and development workshop, for example in its central thematic, which explores the potential use of brain scanning technologies within legal contexts: an emergent issue through the first decade of the twenty-first century with brain scanning used for the first time in both sentencing and convictions.

Musgrave's play was set in the year 2017 (for the 2013 tour). Emerson is 18 years old, has recently lost his mother and has been expelled from school. He finds comfort in free running on the roofs and ledges of the estate where he lives. After a violent reaction to provocation by a local drug dealer, he is charged with grievous bodily harm and faces trial. His girlfriend, Hayley, supports him and they both befriend Miles, another boy at their school who is troubled and socially isolated. During the police investigation Emerson undergoes a brain scan by a forensic investigator, Jeanette, for use as evidence in his trial.[17] In sentencing, he takes the option offered by the judge to choose an experimental deep brain stimulation procedure to control his aggressive impulses rather than a prison sentence.[18] At the end of the play eight years have passed and Hayley, now a lawyer, returns to challenge Jeanette's evidence in another case.

The issues in the play emerge from the relative experiences of the two boys Emerson and Miles and the dialogues between Jeanette and Hayley. Emerson is diagnosed with Antisocial Personality Disorder by Jeanette in the course of the investigation, which provides both mitigation in his trial and opens up the possibility of the medical intervention within the sentence. At the end of the play eight years on, he has not reoffended but he has changed. There are side effects to the technology including not being able to experience the 'buzz' he got from free running, a loss of confidence and a change in his relationship with his girlfriend. Miles displays a number of behaviours that trouble him throughout the play, but he is not diagnosed in terms of any recognised condition. At the end of the play in a monologue he reflects back:

> But then I went to university.
> And I started meeting people, all kinds of people, some of them a bit like me.
> And I realised even though I was different, and I didn't feel like everybody else did… neither did anybody else!
> And I wasn't alone. (Musgrave 2012, 69)

Jeanette defends her work and the move to 'neurojustice'. There have been fewer wrongful convictions; there is reduced violent crime; Emerson and others have been kept out of prison. Hayley questions the notion of a 'normal brain' and draws attention to side effects of diagnosis and treatment and methodological critiques of imaging studies referring to small sample sizes and the homogeneity of control groups (Figure 4.3).

As I have commented on above in relation to *The Gift* and *Dayglo*, the science and technology here is presented as hybrid and open to question. Even Musgrave's script and the staging in performance by the actors, designer Jaimie Todd and director Nigel Townsend do not bring closure or a final frame. As in the TIE tradition, the end of play is not the end of the Theatre of Debate process. After the play has finished the audience are posed a series of questions, listed in Figure 4.4, which they vote on

Figure 4.3 Jeanette (Heather Bleasdale), Emerson (Petrice Jones) and Hayley (Sophie Gerrish) in *Stunted Trees and Broken Bridges*

Q1 Should courts be allowed to use brain scanners as a way of determining whether a defendant is lying or not?

Q2 To what extent do you think Emerson was responsible for his actions?

Q3 Do you agree with the judge's decision to sentence Emerson using deep brain stimulation, a technology that intervenes with the brain, to manage his condition?

Q4 Should neuroscience inform the law?

Q5 How far do you think we should research into neurotechnologies?

Q6 Would you use a technology that intervenes with the brain if you were ill?

Q7 Would you use a technology that intervenes in the brain for non-medical purposes, such as gaming or improving your cognitive skills?

Figure 4.4 Debate questions for *Stunted Trees and Broken Bridges* (Finegold *et al.* 2014)

using electronic handsets. The voting results are shown on a screen and a facilitator from the company poses follow-up questions to the audience inviting volunteers to say how they voted and why. During this process, actors still in character comment on their perspectives on the questions and answer questions from the audience.

This questioning invites a reframing and is generative of different issues partly through the way that it repositions the audience in relation to the narrative of the play. Questions 6 and 7 were asked both before and after the play and seem to position the audience as potential patients who might need to give their consent or consumers who are offered different forms of therapeutic or technological interventions. Questions 1 to 5, which were only asked during the debate after the play, reposition the audience as citizens or 'somatic experts' in Rose's terms. The company collated the votes from across the tour, which reached a total audience of 4,951 across 63 performances.

Some students changed the way they voted in response to questions 6 and 7 when they were posed again after the play. When asked about non-medical uses of technology (question 7) before the play, there were similar proportions of students voting 'no' (41 per cent) and 'yes' (44 per cent) with the rest 'unsure' (15 per cent). After the play, this shifted with an apparent move from 'yes' (33 per cent) to 'no' (53 per cent). Students seemed slightly less inclined to consider the non-medical use of neurotechnologies. In the question about medical interventions (question 6), there seemed to be a move from 'unsure' (34 per cent) before the play to 'no' (34 per cent) after the play with 'yes' shifting only marginally down (44 per cent before, 42 per cent after). However, this rather obscures a greater degree of flux under the surface as records of the patterns of voting showed movement between positions in all directions.

It is not the case here that the ethical judgements made in response to the voting questions were reliant or contingent on specific outcomes of objective (science) learning. Many students voted the same way before and after the play, which also rather belies any idea that the programme created engagement within a completely disengaged group. In this case, students had opinions on the questions and interest in them before the programme, in that they voted yes or no rather than unsure at the start of the event, and could even provide reflection on their position in discussions I observed before the performances. However, some Theatre of Debate projects have worked with objects that were either very new or very remote from everyday experience. In either case, the learning about scientific objects is related to the shifting ethical positions taken by students. The programmes even make links between them, for instance in positions taken by characters in the plays. However, particular understanding rarely determines an ethical position, let alone provides a simple rationale for making a particular choice or voting decision. The uncertainty, contest and shifting terrain is the substance of what students learn. They are also, importantly, learning to be moved and move their own views around in relation to what they know. This is the kind of learning that those advocating changes in the science curriculum at the turn of the millennium seemed to be calling for.

New science and technology often becomes framed in a specific, dominant and fixed way early on in its development and, as discussed earlier, science and technology can become monolithic in the process. An aspect of this phenomenon is what scholars in science and technology have called black-boxing, whereby a concept or device becomes a fixed entity: a black box that cannot be opened for development, reframing, questioning or critique (Pinch 1992). This can have an imbalanced but uncontested influence on alternative ways of knowing. *Stunted Trees and Broken Bridges* highlights how this might happen within the context of neurotechnology. In the play, the character of the judge jokes to the forensic neuroscientist, 'you'll have us all out of a job' (Musgrave 2012, 56), succinctly encapsulating concerns about the growth in the use of neuroscientific evidence in law without sufficient public debate. The shifts in student opinion as a result of watching the play and taking part in the debate suggest that the experience facilitated an unfixing of opinion that might be a result of increased knowledge, a greater appreciation of social implications of technologies or a mixture of the two. It is the very experience of changing one's mind and the process by which students might do this that I suggest underlie its value as pedagogy. The debate questions challenge students to respond speculatively as patients or consumers making choices about their own health or cognitive enhancement and then to think as citizens, voters or jurists in making judgements about broader ethopolitical questions. A further change in frame happens when students see the patterns of their collective voting and discuss why they or others might have voted in such a way or indeed changed their minds as a result of the experience.

Part of the challenge to (political) participation occasioned by life politics is in shifting forms of expertise already discussed and shifting relationships between ways of knowing. In his examination of the emergence of the interdisciplinary field of 'neurolaw' that covers the same terrain as *Stunted Trees and Broken Bridges*, Martin Pickersgill explores knowledge hierarchies and the way knowledge disciplines are imagined:

> *Law* is imagined as first, epistemically subordinate to neuroscience, and second, highly plastic: neuroscience not only should but could enhance legal institutions and processes. At the same time, *science* is imagined as having an intrinsic normativity that demands attention and action, and *scientists* are understood to be key figures needing to be enrolled as part of the assemblage of actors that can and must effect this legal shift. (Pickersgill 2011, 36, italics in original)

Central to the production of this new discipline and epistemic relationship is what Pickersgill, after Jasanoff and Kim, calls a 'socio-technical imaginary'. Disciplinary pedagogies and indeed broader knowledge practices that separate the social from the technical and indeed the social and the technical from the imaginary frustrate the development of individual capacity and the formation of social processes that might offer critical purchase on socio-technical imaginings. This critical purchase, I argue, relies on ethopolitical pedagogies and the development of somatic expertise.

Ethopolitical education and somatic expertise

My use of the concept of framing here has been influenced both by understandings of the way theatre-in-education makes use of different frames to reposition the audience in ways that support learning, as well as the centrality of the idea of issue framing within science and technology studies (Callon 1998; Jackson 2007).[19] As Michel Callon points out, there is a tendency in contemporary technoscience to produce 'hot' situations in which frames cannot contain the interactions within them, producing what he calls 'overflows'. For example, the institutions of bioethics emerging in the 1990s and early 2000s could be regarded as attempts to contain overflows from the pre-existing ways of framing genetics, neuroscience and their new technologies. It is in the overflowing of frames that objects become issues and it is issues that constitute the political, providing the terrain for debate, problems to be contested and mobilising collectives as publics. Institutions, in particular educational ones, have a tendency to bring premature closure or containment through the imposition of particular frames. In this way the heat is dissipated, publics are demobilised, engagement stops and ethopolitics is curtailed.

Ethopolitical education requires processes that sustain engagement, allow the generation of heat, the emergence of issues and the mobilisation of publics. Ethopolitical education needs to be felt as well as thought, relying as it does on the development and valorisation of somatic expertise: the capacity to feel one's way through an issue and sense the emergence or premature closure of issues through the imposition of particular frames. Many of the responses to the Theatre of Debate process that I have heard in discussion with young audience members have mixed references to devices, science and technology with the language of feelings, either commenting directly on their own experience or on the experiences of characters in the drama:

> And the way being told something just because someone's in a white jacket. You get a bit… you panic.
>
> It's about different people, how differently they react to other people and how they can overcome it with the technology these days.[20]

Students here are identifying particular ways technology shapes and is shaped by embodied experience. This includes the way technology might include social or symbolic features such as the authority provided by a white coat. Young people also directly referred to the way the feelings of the characters generated their own felt experience:

> It's like abstract and disturbing, like the way they all move around and stuff like that.
>
> Also, you can almost feel the mood of the characters. It kind of gives out that emotion.

The experience of watching the play is reflected on as embodied, generative of feeling at the same time as the audience members recognised the co-production of this somatic experience with science and technology.

Audience members also showed sensitivity towards the way characters in the plays developed somatic expertise. They commented on how this expertise is created and particular ways it might be used, for instance within frames provided by professional or disciplinary roles:

> [the play] showed clearly, like, how she [Hayley] held in everything about Emerson, like, everything had happened and then taking it on, like, when she was older, when she was a lawyer and stuff. And then she compared it to a different situation with that Sammy.

Sometimes these responses also went beyond what could easily be articulated in discourse:

> We were all talking and then Miles mentioned his mum and everyone went 'oooooh', like we knew he [Emerson] were gonna smack him.

When student audiences responded to voting questions that repositioned them from theatre audience member to potential consumer, citizen or ethicist this shift is not just discursive, it is embodied. Such shift in perspective or indeed the shift by which one changes one opinion on an issue as referred to above is a somatic practice. It involves the exercise and development of somatic expertise.

The development of somatic expertise through Theatre of Debate though has not been restricted to its mostly young audiences but has also been reflected on by artists, scientists and educators. Again this is often in terms that stress the learning as embodied experience and as a challenge to established ways of framing or locating knowledge. The software engineer, Martyn Thomas, who was an expert advisor on two projects Y Touring worked on with the Royal Academy of Engineering, that dealt with privacy and surveillance and electronic patient records respectively, commented in an interview how when attending his first Theatre of Debate workshop he 'experienced a massive culture shock' (Theatre of Debate n.d.). He talks about participating in ice-breaker activities with the group of students involved in that workshop as a 'daunting experience' and remarked that 'it's not the relationship you expect to have when you go into somewhere as, in quotes, an expert'. He goes on to explain the way that the process changed both how he delivered his presentation and the kinds of questions and interactions he had with the young people and others at the workshop. He alludes to the removal or reduction of barriers, that he seems to understand as two-way, that is to say as much a barrier to his understanding of young people's perspectives as their understanding of his expertise.

As I have discussed in Chapter 1 of this book, performance pressure and performance management are manifested in various contexts in which science is performed. This has been particularly the case within school education in the UK, where a succession of reforms since the Education Reform Act in 1988 has attempted to standardise performance measures at the level of the nation, the school and the child and thereby enact particular hierarchies of knowledge and disciplinary boundaries.[21] The period during which Y Touring operated saw increasing expectations for arts and education organisations funded through the state or even from the charitable sector to provide evidence of the efficacy of their practice according to more or less precisely delineated value frameworks. The position of the practice at the interface between different, sometimes contradictory, frames of reference demanded a nimble approach that, as I outlined at the start of this chapter, was resourced by the origins of the company in TIE, health activism and youth work and the early experience of collaborating with health authorities on work within schools.

Over 25 years, the company therefore had to frame and reframe its own practice many times in negotiations with different ways of understanding and valuing learning. Y Touring was consistently able to articulate the value of its performances in a range of terms right up to its last two productions *Stunted Trees and Broken Bridges* and *Hungry* by Sarah Daniels. Since 2014, Townsend has again had to rearticulate the value of work in the terms of the new independent company Theatre of Debate. Figure 4.5 demonstrates the efficacy of the *Stunted Trees and Broken Bridges* programme in terms that might be described as scientific literacy.[22] The sections in the middle of the bars attest to the development of a dimension of somatic expertise that includes the comprehension of socio-scientific discourse.

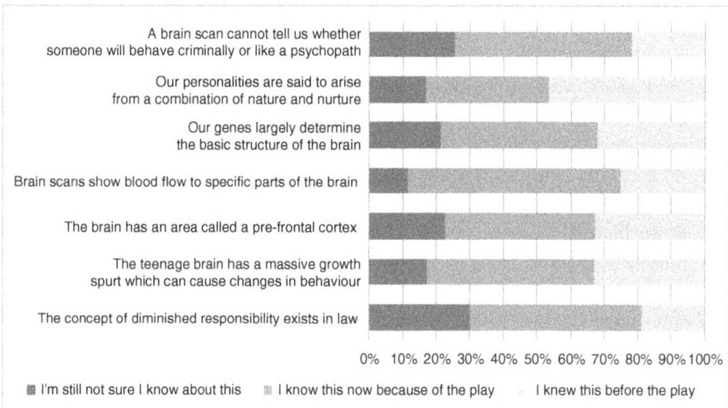

Figure 4.5 Student self-assessment against learning outcomes

That the Theatre of Debate practice could be accounted for in terms that were intelligible to schools and funding authorities and were politically pragmatic does not detract from the extent to which it should be considered as a critical or challenging form of pedagogy for such institutions. Ultimately the closure of Y Touring in 2014 perhaps showed the difficulty of maintaining such political positioning. Central YMCA's concentration on a biopolitics of self-health – of gyms and healthy eating messaging – perhaps finally departed too far from Theatre of Debate's somewhat different approach to ethopolitical education.

Dominant school curricula and pedagogies continue to demarcate knowledge in ways that tend to fragment and thereby limit development of somatic expertise. Critical science educators like Ralph Levinson (2010) have shown how rhetorical commitments to education for scientific citizenship and democratic participation are rarely reflected in curriculum reforms or dominant pedagogy. Many comments from teachers and students prompted by Theatre of Debate have contained implicit or explicit critiques of standardised curriculum or pedagogy. One student memorably remarked to evaluators from People, Science and Policy during the 2004–5 schools tour of *Mind the Gap*: 'the play is not really that relevant to our school work; it is more relevant to society… it gives us the chance to express our opinions in an open and adult way' (People, Science and Policy 2006, 8).

Students' comments about Theatre of Debate have also been reflexive about the way that knowledge is often fragmented within school learning and embodied or social learning is detached and separated from knowledge about science and technology. Such comments reveal the association of particular pedagogies with particular ways of knowing and the association of particular ways of knowing with particular rather limiting affects:

> They gave you a story and showed you what happened. I thought it was going to be all scientific but it had a background to it.
> I thought it was going to be quite boring but it was actually really exciting…
> It's the word science, it makes you think of something where you're doing really hard work.
> Yeah, like it'd be someone standing at the front talking for two straight hours.

The emphasis students have placed on felt responses to narratives about science and technology also suggest the need to value this as learning in a way that is not always properly articulated in theories of critical pedagogy, which focus on dialogue, discourse and deliberation. Theatre of Debate reorients educational theatre around the politics of life itself.

At the same time it offers a radical challenge to science education to remake itself as embodied practice.

Importantly, an ethopolitical education and the building of somatic expertise involves the integration of and negotiation between different ways of knowing and different ways of framing knowledge. The history of how AIDS activists were able to change the drug approval processes and approaches to clinical trials in the USA illustrated how ethopolitics relies on the integration of different ways of knowing, as well as institutional boundary-crossing and the juxtaposition of different ways of framing a problem. Activists drew on their embodied experience as patients and carers. They mobilised communities and founded new collectives. They forged alliances with drug companies, clinicians and scientific researchers. They also developed the capacity to use the language and practices of biomedicine often gained through informal study but sometimes within educational institutions.[23] This process had to be responsive to the emergence of a novel object, HIV/AIDS. Education for ethopolitics requires helping young people develop the somatic expertise to be able to respond to novel objects it might currently be impossible to predict.

Notes

1 Y Touring, while primarily operating as a theatre company, developed partnerships with video production companies and adapted some of their stage plays for television broadcast and/or video/DVD distribution. A filmed adaptation of *The Gift* by Nicola Baldwin was widely distributed in this way. The company were also early adopters of the convention where live theatre is filmed and then broadcast either live or recorded and then screened as live in cinemas and/or streamed online. Filmed performances of five Theatre of Debate productions are available via the Digital Theatre Plus archive: *Mind the Gap*, *Dayglo*, *Stunted Trees and Broken Bridges*, *Hungry* and *People are Messy*. I have used this archive www.digitaltheatreplus.com as a reference source for this chapter as well as having seen live performances on numerous occasions.

2 For a period of time Y Touring ran the venue One KX in Judd St close to King's Cross station in Central London, running programmes for local young people as well as developing their touring work.

3 See the successive editions of Tony Jackson's *Learning through Theatre* for a sense of the history, debates and evolution of the movement (Jackson 1986, 1993; Jackson and Vine 2013).

4 It is discussed briefly in Steve Ball's (1993) chapter of the second edition of *Learning through Theatre*.

5 Mark Harrod (2011) has reflected interestingly on the way the YMCA as a global 'values-based organisation' has held together sometimes competing value frameworks in different ways depending on the location of individual

YMCAs. These reflections resonate with some of the conclusions I come to in this chapter about Y Touring's practices.

6 This proposition relies on histories of modern biology by Foucault (2001) and Canguilhem (1994) which underpin the work by later scholars including Sarah Franklin and Nikolas Rose whom I refer to in this chapter.

7 A list of these Y Touring productions engaging in some way with genetics or genomics might include four plays commissioned for young people to perform between 2003–4 (*Scenes from the Fair* by Jonathan Hall, *Genes 'r' Us* by Rahila Gupta, *Born of Glass* by Rhiannon Tise and *Leap of Faith* by Nicola Baldwin) as well as Theatre of Debate projects such as *Pig in the Middle* by Judy Upton, *Sweet As You Are*, *Making Astronauts* and *Learning to Love the Grey* all by Jonathan Hall and *Nobody Lives Forever* by Judith Johnson.

8 The relationship between the BRCA genes and breast cancer is more complex than the genetics of the single gene disorder, e.g. the FXN mutation that causes Friedreich ataxia featured in *The Gift*. There are many SNPs discovered to date that result in what is termed a faulty BRCA1 or 2 gene.

9 Other Y Touring plays dealing with neuroscience or mental health have included *Mind the Gap* by Abi Bown and *Where's Your Head At?* by Rhiannon Tise.

10 In the 1990s, specific genetic mutations were identified and tests developed for inherited conditions like Huntington's disease and Friedreich ataxia (Bates 2005; Campuzano *et al.* 1996).

11 By 2019 400 conditions were approved for testing by the UK's Human Fertilisation and Embryology Authority (2019).

12 In its tours the dates were adjusted so that the scenes with Ryan, Annie and their mother remained in the present. I am using the dates here from the tour which took place in 2000. It was also made into a video drama in collaboration with Maverick Films/Zenith North/BBC Learning Zone.

13 The character was based on Poly Styrene the lead singer of the punk band X-Ray Spex who died of breast cancer in 2011 (Swash 2011).

14 The gene for sickle cell is not linked to the gene for skin colour but rather to a certain level of resistance to malaria, so is more common in people whose heritage is in areas where malaria is prevalent. Race is used in medicine though as a proxy for an increased likelihood of certain genetic predispositions such as sickle cell, even though many scientists regard the concept of race as meaningless in (genetic) scientific terms (Root 2003). Sickle cell is a single gene disorder like Friedreich ataxia, but the same mutation in the β-globin gene can lead to a variety of symptoms (Frenette and Atweh 2007).

15 Millar and Osborne's research informed the development by the University of York Science Education Group of the OCR *Twenty First Century Science* GCSE qualifications launched in 2006 (University of York Science Education Group n.d.).

16 I worked with Y Touring 2002–9 as a project manager at the Wellcome Trust. I then was part of an evaluation team working with the company 2011–17 on projects created within this period. This chapter reflects from a position influenced by, but after, these relationships.

17 fMRI was used in sentencing evidence in a murder case in the USA in 2009 and in a murder conviction in India in 2008 (Miller 2009; Saini 2009).
18 The use of a brain implant in the story is more speculative within this (legal) context although it is now widely used in treatment for particular neurological conditions, e.g. Parkinson's (Wahlberg 2014).
19 These uses of the idea share a starting point in the work of Erving Goffman. Goffman (1986) showed how particular frames established the rules for social conduct within a certain context, setting up a boundary but not cutting the context off from the external world. Institutions establish frames that attempt to contain interactions between subjects and objects in order to perform their functions in education, health, criminal justice, art or other social fields.
20 Quotations from audience members in this section are drawn from focus groups after the performances held as part of the evaluation research on *Stunted Trees and Broken Bridges* that I conducted with Jo Hutchinson (University of Derby) and Peter Finegold (Isinglass Consultancy).
21 National performance is often evaluated via the Programme for International Student Assessment (PISA) tests established in 1997 by the OECD. Individual schools have been assessed since 1992 via the Office for Standards in Education (OFSTED) and school league tables. Individual children are assessed in nationally administered qualifications and via Standard Assessment Tests (SATs).
22 While these look like curriculum-based learning outcomes, they are not. They were developed retrospectively on the basis of the playscript rather than prior to its development as one, but not an exclusive, framework for evaluation. These learning outcomes can be related, but are not integral to, the school science curriculum. This process in itself reflects the way Theatre of Debate flexibly accommodated itself to the discourses and frameworks of its stakeholders while not allowing these frameworks to overdetermine the practice.
23 This history is documented in a number of texts (Epstein 1996; Gould 2009). It is discussed specifically in relation to scientific expertise by Harry Collins and Trevor Pinch (1998).

5

Commoning sense: translating globalised knowledge in performance

Some time in 2006 when I worked at the Wellcome Trust, I remember being contacted by Matthew Walters, a young creative producer working in the Handsworth area of Birmingham (UK). He reported on a successful first meeting between a group of grime MCs and dermatologists at the local hospital. I remember this moment as, for some reason, the meeting between grime MCs and dermatologists seemed to exemplify a pioneering and important act of boundary-crossing. It was notable or unusual because of the apparent gulf between dermatology researchers or clinicians and the young grime artists. These are not questions of physical distance as such – it is a 30-minute bus ride from the hospital to Handsworth and many people from the area will be treated at the hospital. What constitutes this gulf or boundary then? Why might it be a problem in political terms? Is it helpful even to think the problem in spatial terms like gulf, boundary or gap? How might creative practices respond to such boundaries by reconceptualising them or even bridging them?

Influential accounts of difference and equity in relation to scientific knowledge have tended to think this difference economically rather than spatially in terms of 'science capital' (Archer *et al.* 2012). This theoretical model follows Pierre Bourdieu's sociological accounts (1977, 1984) of the way that power circulates within and through culture. Bourdieu's account of cultural or symbolic capital relies on an analogy and association between power within a particular cultural context and economic capital. Archer's work on science capital within education is important in showing how scientific knowledge and its accumulation – sometimes understood to be egalitarian, transparent and apart from culture – is similarly affected by socio-economic factors to other kinds of symbolic capital. This work has significant implications for pedagogy and support for careers in science, technology, engineering and maths. It highlights a political problem in terms of the way hierarchies and inequities are reproduced through cultural practices, including scientific knowledge-making.

However, what this conceptual approach risks rehearsing is the premise of the deficit model in its various incarnations. The notion of a deficit, whether conceived in terms of capital, engagement or knowledge, presupposes a lack, or absence. It neglects what might be there and conceals value judgements implicit in such a presupposition. Deficit models are premised on a rather static conception of what constitutes knowledge or engagement, failing to recognise existing resources or the exclusion created by dominant ways of framing an issue. As feminist scholars of politics have pointed out, difference can become a resource for political communication and for reimagining the practices of politics (Young 2002). Scholars that I have referred to elsewhere in this book, working within postcolonial and science studies frameworks, have proposed notions such as cognitive or epistemological justice to address the devaluation of different modes of knowledge production (Leach *et al.* 2005; Santos 2014; Visvanathan 2003). These ideas also help explain the links between inequities in the value of knowledges and other kinds of inequity. Boaventura de Sousa Santos has argued influentially that 'there is no global social justice without global cognitive justice' (2014, 42). Santos is concerned with global power dynamics between North and South and with the legacies of colonialism; however, such power dynamics and legacies also play out in various ways within local contexts in all parts of the world.

The practices I discuss in this chapter all happened in the UK but show how knowledge is dispersed and differentiated across space that is produced locally and globally along what the geographer Doreen Massey called the 'power-geometries of time-space' (1999, 27). Arts projects discussed in this chapter took place in London, Manchester and Birmingham, three big cities in the UK, and yet they all drew on various knowledges that were produced transnationally out of global networks of scientific researchers, through hybrid media and artistic forms, across public, private and third sectors, between institutions and informal groups, as a function of urban geography and the geographies of diaspora. These knowledges are not valued equally. They are subject to Massey's (1999) 'power-geometries' whereby globalisation, colonialism and neo-colonialism shape and are shaped by relations of co-production. Santos (2014) draws particular attention to the way knowledges produced by communities in some parts of the Global South, such as indigenous peoples in South America, India or Africa, are both ignored and expropriated by powerful actors – states and corporations – predominately from the Global North. A feature of this exploitation is the collapsing of space into time, so the means by which indigenous communities or others from the South are imagined as temporally 'behind' the urban centres of the North (Massey 2005, 7). This involves both a corruption of historical narratives and a particular epistemological move through which knowledge production is seen to 'progress' from vernacular or traditional processes to the institutions of modern science.[1] There is much work to be done to explore how arts practices might support the negotiation between different ways of knowing, in particular between science and traditional knowledges, in different parts of the world. In this chapter, however, I am focussing on how such global politics of knowledge play out within the UK's big cities and how arts practices might intervene in such politics.

Walters' meeting in Birmingham took place in the context of a project called *Colour Coded*. It was part of a process he had devised to explore skin colour, to engage with debates about the use of skin bleaching products among different groups of young people he knew in the Handsworth area. Skin, skin colour and a number of legal and illegal cosmetic products provided objects of interest shared by the different groups. They were all engaged with and through their skin. Everyone knew something about their skin and its pigmentation and had ways of making sense of it as feeling, meaning or source of knowledge. What appeared in the course of the collaboration therefore was not addressing a deficit as such but the articulation of difference. There were significant differences between the way different groups sensed and expressed what they knew. There were also significant questions of access for different groups to each other's sense of and knowledge about skin. Such questions of access did of course have partly to do with inequalities and social value. However, I argue in this chapter that practical, creative responses to knowledge disparities might be better conceived not in terms of redistribution or deficit reduction, but in terms of practices of translation. The notion of translation crucially encompasses discourse and space as well as gesturing towards ways in which space and discourse are intertwined in the production of common sense or knowledge held in common. In this respect, the project and this way of thinking about it resist tendencies to conceive of and produce relations in terms of deficit or to collapse space into time.

To think about knowledge co-production in terms of translation is to take into account the diversity of knowledges, not necessarily equating them, but recognising them and thereby valuing encounters between different ways of knowing. If the practices discussed in this book can be thought about as commoning sense, the 'sense' that is shared here is not common sense conceived in opposition to science, neither is it a unified and monolithic common sense. Rather it is sense commoned by the way it accommodates contestation, different experiences, and dynamic, evolving collectives and understanding. In the previous chapter, I charted the development of Theatre of Debate, a particular approach to ethopolitical education that validates somatic expertise through a form of co-productive dramaturgy. This approach has been carefully evolved, building networks between schools, artists and scientists through narrative. In this chapter I look at the way artistic performance practices open up a range of translation possibilities, including and beyond narrative, through flexible and exploratory play with language and space. The practices discussed here bring together forms of co-productive dramaturgy in ways that emerge from diverse groups, issues and feelings for and about science. As such, they connect many of the theatrical sense-making practices I have discussed in earlier chapters. Opening up such diverse forms of science in performance is necessary in developing more just or equitable knowledge-making processes and responding to the challenge of cognitive justice that Santos outlines.

To some extent, such play is disruptive of science as a monolithic or dominant sense-making practice. As Santos has argued, 'modern science built itself against common sense', if common sense is conceived in mythical or mystical terms (Santos 2014, 157). Such a conception can give rise to the expropriations and dismissal of other ways of knowing. However, the practices discussed in this chapter all involve collaborations and encounters between some participants seen as holding authorised specialised scientific knowledge and other participants who do not. These encounters are not, by and large, easily characterised in simply oppositional or antagonistic terms. Rather, they accommodate and provide a form that can hold different ways of knowing rather than order knowledge in hierarchical terms or attempt to prematurely synthesise perspectives. In the artworks discussed here, it is the aesthetic and social form, the process of commoning sense, that is created by and constitutes the encounter. The encounter is in itself the form of the knowledge. Santos proposes solidarity as a form of knowledge and contrasts this with hegemonic knowledge defined by its capacity to create order:

> Solidarity as a form of knowledge is the recognition of the others both as an equal, whenever differences make her or him inferior, and as different, whenever equality jeopardizes his or her identity. (2014, 156)

The artistic forms discussed in this chapter use different social processes that enable recognition of diverse ways of knowing. They also create aesthetic forms that facilitate the co-location of knowledges in ways that resist premature ordering. In each case an object provides a starting point: skin, breast cancer, khat. As I have discussed in the previous chapter, these were all objects that in some way were 'hot' for the diverse groups who assembled in relation to them and were thereby productive of issues. However, again as in the previous chapter, the artistic practices I discuss here departed from mainstream modes of science communication or engagement in the way they produced and framed issues and notably in the ways they formed alliances through generating solidarity. This involved translating across difference that can be conceived spatially and discursively.

In the first section I explore how narratives about breast or ovarian cancer and inheritance have been subject to a series of public translations within recent history. I discuss two theatrical projects which offer somewhat different perspectives to the more dominant stories. In some ways, these practices bear a close relationship to the ethopolitical dramaturgy of Theatre of Debate. They validated both somatic expertise and scientific knowledge and were able to hold these together even if there were apparent tensions between them. They are examples of biosocial theatre in the sense that they emerge out of and contribute to the formation of what Rabinow (2002) calls 'biosociality' – alliances between patients, scientists and others that I have referred to in relationship to genetic conditions and HIV/AIDS in particular. In the second section, I will discuss the translational work of *Khat Out of Sight Out of Mind*, another example of biosocial theatre that operated within the particular context of the Somali diaspora in East London and enabled an alliance between medical professionals and a broader local community through an artistic form involving linguistic translations of a variety of types. In the last section, I return to the dermatologists and the grime MCs, showing how they used media and performance to map knowledge onto urban space. I also explore how the radical discursive differences between the language of grime and the language of dermatology became a resource to highlight the incompleteness of both as ways of knowing about the skin.

In each of these cases, theatrical modes of representation, so particular configurations of 'time-space' as Massey (2005, 177) puts it, with their own power dynamics responded to other more dominant modes of representation, for example the theatrical performances of reworked cancer stories as narrated in consultations with clinicians but also as told by celebrities via the mass media. *Khat Out of Sight Out of Mind* was in part a response to official expert reports intended to inform government policy. *Colour Coded* was a reaction to advertising for particular cosmetic products as well as to scientific representations of skin health as found in academic journals. The artistic projects all reopened apparently premature closures performed by dominant representations of their matters of concern. They did this by finding forms of translation that enabled the co-existence of multiple stories and, to some extent, started to mitigate the unequal 'power-geometries' that shape time-space (Massey 2005, 82).

Translating BRCA: problematic narratives

Starting to tell the story – where and when to start to talk – about cancer is itself a key element of the process of translation. At the same time, the identification of the characters in the story determines or is determined by the starting point of the narrative. Popular narratives about cancer in drama or news media tend to tell the story from the point of view of a particular patient or a medical professional. Diagnosis, finding or detecting a tumour, either starts the narrative or is a central moment within the dramatic structure. Scientific cancer narratives may take a historical perspective starting from particular discoveries of causes or treatments for particular cancers, such as Sidney Farber *et al.*'s (1948) pioneering work on chemotherapy for leukaemia.[2]

In his sociology of translation, Michel Callon (1986, 204) describes the moment when actors are determined and their identities defined as 'problematisation'. Part of Callon's symmetrical approach, which fed into the development of actor-network theory along with Bruno Latour, John Law and others, allows agency to both human and nonhuman actors. So translation can involve competition between actors to determine and define each other in a way that makes themselves 'an obligatory passage point' in a network of relations (1986, 205). Molecularisation and the emergence of genetics and genomics, which I have referred to already, enables a reproblematisation to the cellular and then sub-cellular levels that recasts the protagonists of cancer narratives as particular genes or networks between genetic and environmental materials. This potentially enables translations to be made between such problematisation at the molecular level and treatments that also operate at the molecular level. However, such translation tends to separate such a way of knowing about cancer from popular, cultural ways of knowing about it. It also relies on translations from molecular problems into problems of global and local economics and indeed the problems of research and clinical institutions. These translations can become both opaque and reduced within dominant narratives.

A story, or indeed a(n as yet unmade) Hollywood film, of breast cancer genetics could reasonably begin in 1990, when the US scientist Mary-Claire King and her team made a link between family history of early onset breast cancer and DNA polymorphisms at a particular location on chromosome 17 (Hall *et al.* 1990). This became known as the BRCA1 gene. In this problematisation, a triangular network is formed between cancer tumours, so the symptoms of cancer such as a felt or visualised lump, families who across generations experience cancer symptoms, and researchers like King attempting to identify causes of these symptoms. Generations of researchers have defined cancer as a set of symptoms identifying agents initially at the level of organs (breast, bowel, lung, etc.), then at a cellular level and then in molecular terms. Researchers, such as King, go on to define its transmission genealogically, identifying not just individuals but families as agents in this transmission. Such a problematisation establishes King and her colleagues, or at least their theoretical frameworks or concepts, as an 'obligatory passage point' in understanding certain types of cancer. The scientists themselves are ultimately displaced after further translations by various technologies including that of genetic screening for BRCA1 and 2. The story told in this way foregrounds a moment of discovery or revelation: a significant shift in popular scientific understandings of cancer which had been seen primarily as the result of somatic mutations in previously healthy inherited DNA caused by factors such as smoking, exposure to radiation, diet or aging. Within a popular history of science, the collective scientific endeavour that King and a number of others were involved in has also been narrated as a race with a complex social context of competition and collaboration reaching a denouement in a published paper and accepted results. The tendency in the historical record is to narrate at molar scale while the scientific record concentrates on the molecular.

In Callon's terms, the displacement that happens here occurs through the next 'moment of translation' that he calls 'interessement' (Callon 1986, 206). This is a process by which allies are recruited and various devices are used to stabilise a particular problematisation. In King's case, allies included other colleagues who needed to be persuaded of her theories and research funders, and might also include the genes and the variants themselves. The devices used might be genealogical charts but also genome sequencing technologies and the various scientific approaches that have led to the identification of variants in BRCA1 and later on BRCA2 genes that increase the risk of breast or ovarian cancer. Of course, these translation processes will involve various overtly performative activities by the agents, in that the mutations are made to perform within various performance forms. These might include scientific conference papers and laboratory experiments. In these scientific performances, animal actors, notably the oncomouse, as well as human cell lines sometimes theatrically stand in for human actors making the mutations appear and enabling particular kinds of representation.[3]

All these moments of translation are moments in which competition and contestation, multiplying stories, can and do take place. This contestation can often take place through performance. Even when focussing on the way one individual (human) actor, in this case the eminent scientist Mary-Claire King, has employed performance forms within translation, it is possible to find a number of examples where different modes of theatricality change the problematisation and change the story. King herself has subverted what was apparently her own problematisation, in a theatrical version of the BRCA1 story as told by her at the World Science Festival in 2014 when she put herself on stage a bit like the dancing roboticists of Chapter 2. 'The week of April Fools' day of 1981 began badly', she starts. The story both is and is not about her discovery of BRCA1. It is a personal account in which her husband leaves her at a key moment in her career. She is left with her young child on the eve of a crucial interview for a research grant that we gradually realise was 'the beginning of the project that became BRCA1'. The device of the story interests the audience theatrically in establishing its stakes and jeopardy. If she had not gone to the interview, perhaps her research never would have happened. BRCA1 might never have existed. Her mother arrives to help but on discovering the breakdown of King's marriage leaves again in distress. However, King is helped on her way by the benevolent interventions of two older men – a friend and mentor and, bizarrely, the legendary baseball star Joe Di Maggio – who look after her daughter as she makes it to her interview and wins the funding for her research.

Through the story King re-problematises the production of BRCA1, displacing some of the agents that appeared central to the more orthodox historical account. The problem is no longer one in which cancer symptoms, researchers and patients are the network. She does this partly by changing where the story starts – some time before she actually starts the crucial phase of her research. She also does this by drawing on the dramaturgical device of the quest narrative, perhaps inspired by The Moth, the New York-based storytelling organisation and producers of the event, who expect their storytellers to employ particular conventions and indeed coach them in these (The Moth n.d.). King is still the obligatory passage point in Callon's terms or the hero or protagonist in narrative terms. However, her interested allies who help stabilise the problem now become Di Maggio and her mentor along with perhaps an airline and some research funders. In fact this performance may even have displaced cancer or BRCA1 to such an extent that the problem has now become the production of scientific knowledge by women. Of course, the particular theatrical context of this translation requires some knowledge of the one I previously outlined and gives it a bit of a push rather than really destabilises it. Thinking about translation theatrically, as Callon does, makes better sense of what is happening here than more purely linguistic ideas, metaphor for instance, as it takes into account who the actors are, what other roles they have played and the way issues are framed.

Of course scientists are not the only actors who become or create obligatory passage points through the way problems are posed. Scientists like Mary-Claire King actually relied on a translation of cancer within the USA and more broadly such that from the 1970s and 1980s it became a *bio*medical rather than medical problem. Biomedical scientists were therefore 'enrolled', in Callon's terms, as actors within this translation process (1986, 211). Key agents in this translation process in addition to the scientists themselves were the biosocial networks of patients, patient groups, drug companies and clinicians who effectively destabilised the earlier consensus, what Maren Klawiter (2008) has called cancer's 'medical regime', in which the hospital consultant was more central in defining the problem. Within the biomedical translation the doctor and her/his diagnostic and treatment repertoire is partly displaced by a variety of technologies including new drugs as well as technologies of screening that now include genetic screening. Enrolment, according to Callon, can involve various levels of 'physical violence… seduction, transaction, and consent', all of which might be found within the context of cancer histories (1986, 214).

Cancer translations have involved a variety of more or less overtly theatrical devices, notably dramas in various media and the performance activism of patient movements and indeed individual patients.[4] In a *New York Times* article, the actor and director Angelina Jolie (2013) announced that following her own positive test for a 'faulty' BRCA1 gene she had opted for a preventative double mastectomy. Jolie's narrative translation of her experience in the article relies on King's problematisation stating: 'the truth is I carry a "faulty" gene, BRCA1, which sharply increases my risk of developing breast cancer and ovarian cancer.' However, it goes beyond King's problematisation. Jolie's article starts with a death, that of her mother from cancer at 56, and proceeds towards her own genetic status only in the course of reflecting on how she struggles to explain her mother's illness to her own children. BRCA1 and its mutations are a fact, a 'reality' as she puts it, that offer a retrospective account of a story already told albeit with its own consequences and implications for her, and, although she does not directly refer to this, also for her children. For her, that BRCA1 mutations increase cancer risk is uncontested – a fact or reality – and beyond doubt: this problem has been closed.

Jolie now 'mobilises' – Callon's fourth 'moment of translation' (1986, 214) – genetics researchers and the faulty genes themselves, in addition to her late mother and her children, as allies in an overtly stated campaign so that in her words 'other women can benefit from my experience' by undergoing genetic screening and making choices about treatment, for instance by preventative mastectomy, as she has done. Callon points out how, in successful mobilisation, 'spokespersons' acquire authority in silencing those actors, in this case humans and nonhumans, in whose name they speak. Jolie gains authority by speaking on behalf of a number of populations – including the allies listed above – whom she makes mobile in her article. In doing this she draws on her considerable agency within overlapping networks as mother, daughter, patient, or, in the new biosocial vocabulary, a cancer previvor, but also as professional performer and media celebrity.[5]

Jolie's translation capacities are of course exceptional in various ways. James Thompson has criticised the role of celebrity activists who presume to speak for humanity from positions of exceptional privilege. Drawing on Agamben (1998), Thompson argues that celebrities, as 'super lives', unlike those 'bare lives' for whom they speak, have the capacity to 'interrupt' time and also to traverse space in exceptional ways (Thompson 2014, 82). Jolie's access to screening, treatment and advice at the specialist Pink Lotus Breast Center in Beverley Hills, despite their apparently inclusive treatment and screening ethos, was no doubt exceptional particularly within the insurance-dominated and therefore highly unequal US health system. As Jolie points out herself, most breast cancer deaths are in low- and middle-income countries and screening costs more than $3,000 in the USA. Her power to generate publicity through the *New York Times* both grants her exceptional authority and global reach for her story. This story has had extraordinary effects. Researchers in the USA and the UK have shown significant increases in the numbers of women requesting BRCA tests in the time after the publication of the story – what some have called a 'Jolie effect' (Desai and Jena 2016; Press Association 2014). This phenomenon may provide some women with the choice that Jolie was able to make, her stated aim in the article, although the US research suggested that women who accessed the test were not necessarily those at the most risk. The broader effects are probably even more ambiguous, however. Commentators and critics have critically discussed the influence her problem posing may have had, claiming that her exceptional capacities of mobilisation draw focus away from breast cancer causes that are not genetic, other treatment experiences, or indeed other conditions that have not been spot-lit in the same way (Tetteh 2019).

Other commentators have criticised the translations performed by patient narratives, which might include Jolie's, for the way they actually ignore new ideas about diversity that emerge from genetics and translate new problems within the frame of existing narratives – closing down rather than opening up diversity and the multiplication of stories. Diane Herndl summarises her critique of particular examples from memoirs and TV documentary in these terms:

> In these works, the courtship and child-bearing narrative of the individual woman comes to overwhelm a larger narrative about the intellectual commons and the common weal. To that extent, the new stories become, in fact, very old ones. They erase the new narrative of deleterious mutation with the old narrative of courtship, child-bearing, and maintaining an appropriately fit and feminine body. We need to examine this narrative erasure more closely. (2014, 36)

As Herndl points out, there is a risk that narratives, however therapeutic they might be for individual writers and readers, can reinforce an existing and limited common sense. The artistic and political challenge then is to common sense in new ways that take into account patterns of cognitive and social injustice. It has not been my intention so far in this section to pass ethical or political judgement on any of the translations I have described. I rather wanted to develop the idea that Callon's analytical framework might help to show how theatre and performance practices become involved in co-production and indeed how theatre or performance theory might contribute to understanding co-production and the translation of cancer. As the work by Thompson, Herndl and Nye shows, significant ethical and political questions are raised at all moments of translation.

Two theatrical examples from the UK attempted to translate hereditary breast cancer (BRCA1/2) in the wake of Jolie's article and addressed these questions in slightly different ways. *Goodstock* by Olivia Hirst was produced and toured in 2015 by Lost Watch Theatre Company, a London-based company making original contemporary theatre formed by Rianna Dearden, Olivia Hirst and Agnes Wild in 2013. *Genesis* was produced and toured in 2016 by Forward Theatre Project. Forward Theatre Project was founded in 2009 by a collective of directors, writers and designers from different parts of the UK who make original theatre out of long-term residencies in particular communities. *Genesis* was developed with the Prevent Breast Cancer charity in South Manchester and communities in the Manchester area, but the company has developed projects in Scarborough, York, Doncaster, Derby and elsewhere in the UK.

Goodstock explores Hirst's own experience of living with a positive BRCA1 test result and the awareness of the impact of cancer on her own family. Hirst plays herself in the autobiographical piece that proceeds from personal experience, in the same way as Jolie's article, assembling a similar network of actors. It features previous generations – including her grandmother, mother and aunt (played by other actors) – and engages with narratives of courtship (her relationship with her boyfriend), child-bearing (her research into IVF) and the female body. This is succinctly summarised in the preface and promotional text used:

> I'm 26.
> Three women in my family have a cancer-causing genetic mutation.*
> One of them is old, one is me and the other is dead.
> In four years' time, I have some options.**
> Basically this is a play written about my relatives without their permission.
> They won't mind.***
> *BRCA1.
> **Involves a scalpel.
> ***I hope.

What is striking is the way this text shows reflexivity towards the theatrical medium as translation. Hirst prefaces her performance both with a summary of her problematisation, an outline of her enrolment and mobilisations, and indeed a perhaps tongue-in-cheek reflection on the ethics of this mobilisation. Unlike Jolie's article, there is no particular sense of a campaign in favour of screening. Lost Watch has been invited to perform the piece several times within the context of medical education. This is presumably not for its explanations of cancer, via a comically extended cake metaphor sketched on the back of a cello, or its explanations of surgery, drawn on a transparent raincoat, but rather for the way that it articulates relationships between BRCA1, testing and surgical technologies, families and so on. Hirst, having gone through the passage point of the genetic test, shows how she is entangled in networks of relationships with other humans living and dead, but also now inevitably with organs, cells, molecules and a range of medical technologies. The piece is reflective and reflexive, tracing the networks produced by and productive of cancer genetics. At 26 years old, Hirst had not yet had to make a final decision about preventative options. The ending is therefore left open by the piece, although Hirst seems clear at the end of the performance that she has decided to have the operation. As much as it involves storytelling, in some ways, given its lack of resolution, *Goodstock* is a map of the network rather than a conventional narrative. It stages a process of Hirst locating herself within her networks. It is temporally positioned in Hirst's present and held there by her pasts and her possible futures.

The project that led to *Genesis* was originally conceived by another performer, the actor Morag Siller, who sadly died of breast cancer in 2016. Siller had become a patron of the Prevent Breast Cancer charity and proposed the idea of making a play to the artistic director of Forward Theatre Project, Charlotte Bennett. Unlike Jolie or Hirst, at this point Siller had actually been diagnosed with cancer. Prevent Breast Cancer works in cancer genetics and prevention rather than treatments, so to some extent the project was not responding to Siller's own personal story, but rather as a response to her work with the charity. Charlotte Bennett collaborated with the writer Frazer Flintham to create a fictional script based on discussions with patients, clinicians and researchers mostly accessed via the charity. She described to me in an interview how:

> [w]e met big groups of patients, we met patients who have had preventative surgery, who were thinking about it, who had decided they didn't want to know, who had breast cancer. We met doctors, surgeons, geneticists, dietitians, nurses, a huge number of people, we didn't interview them really, we chatted with them about what they did.

This process showed Bennett and Flintham, relative newcomers as they were to these networks, the levels of contestation that still remain about problematisation and stabilisation of cancer even when clinical accounts may seem stabilised. Bennett talks about feeling pulled about by actors within the networks:

> [t]he other thing we found with Prevent Breast Cancer, which was really interesting, was how much conflicting information we got from within the scientific team, so the dietician would come in and say, 'diet, it's all about diet, nothing to do with genetics, it's all about diet'. Then the geneticists would be like, 'diet is a tiny thing, it's all genetics'; we've felt we'd been pulled in every direction.

These feelings of being pulled about, as well as the emergence of the networks themselves, are explored in Flintham's play and Bennett's production through theatrical strategies including the dramaturgy and scenography of the performance as well as additional programmes of activities they organised. The presentation of these feelings in performance was also crucial to members of the audience who reflected on what the play offered through post-show discussions.

The central character in *Genesis* is a cancer genetics researcher, Rachel, who in the course of the play herself tests positive for a BRCA2 mutation. Like Jolie, she is clear about her choices and that preventative surgery is the right decision for her. However, the research process behind *Genesis* is reflected in the way that this narrative is complicated by the exchanges Rachel has with other characters, principally her daughter and her friend. Her initial certainty about her choices turns out to have complex implications beyond herself. Rachel's work as a cancer researcher figures her as two actors rather than one (a researcher like King and a 'previvor' like Jolie). At the start of the play we see her in the former role counselling a patient. At this point she is clear in the way she assembles a network, enrols and mobilises human and nonhuman actors. This becomes increasingly hard for her to sustain through the play. She is not just re-enrolled by the BRCA2 test but also in the competing versions of the problem articulated by her work friend and her daughter. Decisions that seemed clear for Rachel at the start are opened up and left open: as with *Goodstock* there is no neat closure.

Both pieces are revealing in that they show how the BRCA testing technologies and associated knowledge produce social relations that become central to shaping the problem. Both also offer glimpses of the social relations that have produced this genetic knowledge and the technologies, reflecting on certain ambivalences in relation to the knowledge and technology and the impossibility of unknowing or unmaking it. In *Genesis*, Rachel initially cannot understand why people would not want to take the test and know their results. Gradually this perspective comes into view, particularly as Rachel's test results begin to have implications for her teenage daughter. Even though she never changes her own decision, she feels increasingly trapped in the net, symbolised in the production by the set designed by Lydia Denno. In the course of the play, sections of the modular floor are pulled up to form boxes containing structures reminiscent of scientists' molecular models with lights as atoms. These structures are both elegant – a term often used by scientists to describe a model, method, solution, theory or formula – and also inhibit the movements of the actors. At points, Rachel seems hemmed in by them or the molecular knowledge they represent. At the end she holds her daughter's hand as she steps gingerly around the edge of the stage over the boxes (Figure 5.1).

Hirst in *Goodstock* feels obligated to take the test when she knows about it. It has, for her, seemingly become an 'obligatory passage point', to again cite Callon. She says 'I can't plead ignorance because I'm not. I'm not ignorant. I got tested because I already know all about BRCA1' (Hirst 2014, 29).[6] By the end of her piece she seems reassured, and thereby

Figure 5.1 Rachel (Helen Bradbury) in *Genesis*

reassures the audience, by her surgical options. 'I'm going to be ok', she says (Hirst 2014, 29). However, the performance piece itself stages the complex, intellectual and emotional process she has to go through while grappling with the sense of obligation to have the test and in coming to terms with her provisional decision to have the surgery. Hirst carefully and honestly unpacks her understanding of the way the problem is framed for her and has been represented to her by doctors, family and boyfriend, disentangles this explanation from her family history and processes the impact it has on her relationships. She does this through a combination of monologues by her or by her grandmother Elizabeth, played by Illona Linthwaite, speaking straight to the audience, and dialogue between her or her grandmother with other significant figures played by Rianna Dearden. The dialogues give a sense of how her evolving understanding about her status is shaped by conversations with doctors and also by her current relationships and relationships with and between members of other generations: her grandmother, father, mother and aunt. There are a number of stories here that are held together within theatrical time-space in a way that keeps the future open. It is not an account of how she came to a particular decision, neither is it a debate about approaches to diagnosis or treatment. Her often sharply humorous reflections give a complex emotional intensity including anger, sadness, optimism and anxiety to the account. This embodied understanding, her somatic expertise, presumably evolves with every performance as she repeatedly enacts and re-enacts the conversations with her fellow performers and new audiences.

Interestingly, the translation practices of making the performances themselves constituted various emergent biosocial forms that configured networks in particular ways. *Goodstock* has become incorporated within medical education curricula in universities, attesting to the way such courses are increasingly valuing somatic expertise and theatre performance as a way of accessing it. *Genesis* grew out of, and also built links within and with Prevent Breast Cancer. Alongside the performance of the play, *Genesis* as a project included a number of other activities such as an exhibition of photographs by photographer Ailsa Bee of men and women who had had mastectomies 'to celebrate bodies and survival', a 'pamper' day for a group of people who had had breast cancer, a flash mob created with a choir, post-show discussions and workshops in schools. Bennett tried to account for this proliferation of social forms in terms of awareness-raising:

> [w]e wanted to create awareness with it, it's not enough to make a play and say isn't that great, I really wanted to draw up an engagement programme… there were so many things that we had captured along the way that we didn't manage to put in the play that I really wanted to catch.

However, the open-endedness of the play and indeed the diversity of the different forms of engagement suggested an awareness of significant diversity between the publics whose awareness they wanted to raise and a range of different objectives for the overarching project. It also resonates with the recurring theme in this book that theatre and performance forms provide a repertoire for holding together different ways of knowing, in Bennett's terms capturing or catching things that cannot be contained within the narrative of a play but which might also be absent from conventional forms of science communication.

It is as if their research process prompted the need for further iterations of social form and sense-making practices – a search for ways of disassembling and reassembling networks. Bennett has talked about the kinds of things audience members reflected on in post-show discussions:

> We had a girl who came, who was 18, maybe 20, to the show, and she said in the post-show discussion, she was sat with her mum and dad, 'my mum wants me to find out and my dad doesn't want me to find out and you've just shown them exactly what's going through my head'.

The theatre piece thus contributes to a commoning of sense within a family, even if the perspectives shared do not make decision-making any easier. For Hirst as performance-maker and for this audience member, theatre offered both a means of holding together networks including conflicting modes of understanding as well as a vocabulary for understanding their production. It gave form to thoughts and feelings difficult to disentangle within complex and difficult personal experience.[7]

Translating khat

The translation capacities of Mary-Claire King and Angelina Jolie enabled the rapid mobilisation and stabilisation of a particular cancer narrative that *Genesis* and *Goodstock* also attempted to come to terms with on the other side of the Atlantic. Both these projects connected biomedical expertise with the somatic expertise of particular groups and individuals. Sometimes translations are destabilised by the way knowledge is unequally dispersed globally across diasporas and within multicultural cities. This section explores a project that also involved a collaborative theatrical re-problematisation where the development of biosocial alliances was made difficult by significant social, linguistic and cultural barriers that intersected with and to some extent correlated with different ways of knowing.

Khat Out of Sight Out of Mind, a project developed in 2006, attempted to build networks around the hot issue of khat-chewing and mental health within the East African community, predominately of Somali origin, in East London. These networks spanned differences within those of Somali origin that had to do with gender and generation as well as other factors. They also aimed to connect medical professionals caring for local communities with a range of members of the Somali community. In this case, the need for translation as a response to inequalities and disparities among the diverse communities of East London was quite widely recognised (McCrone *et al.* 2005; Mohamud *et al.* 2002). The conflict in Somalia and the relatively rapid dispersal of Somalis from East Africa to various other parts of the world were obvious causes for both the disproportionate level of ill health of diaspora communities and indeed the difficulties such communities experienced in accessing local services for care and treatment. The rapidity of the forced migration also presented challenges for cultural translations that necessarily take place as people and their networks are relocated.

Between 1985 and 2005, Somalia became 'a truly globalized nation' according to the political scientist Ken Menkhaus (2008, 188). Somalia's globalisation is a consequence of decades of civil war, a botched international intervention, and patterns of coerced migration within and out of the Horn of Africa region. Many people fled the country in the 1990s moving to Kenya, Europe, North America or other parts of the world. However, Menkhaus is not just referring to dispersal. Somalia's globalisation is constituted by the movement of people, money, power, commodities and ideas backwards and forwards between Somalia and the rest of world and across a shifting Somali diaspora. Somalis have not been able to rely on institutions of a functioning nation-state and have had to develop mobile and dynamic forms such as the form of 'underground banking' known as *hawala* used to send remittances back to family in Somalia from those working overseas (2008, 192). At the same time, Somalis, whether living in Kenya or Canada, have had to negotiate the institutions of their countries of residence or, in many cases, new countries of citizenship, reconciling their relationships to both these places and the places of their birth or heritage. Experiences of Somalis in this way are particular but also exemplary of the experiences of many others with lives that have been globalised according to particular time-space power-geometries, to use Massey's (2005) expression again. The globalising experience for many Somalis has been, in the drily witty understatement of one Somali poet and scholar I spoke to, one of 'some disruption'. This disruption has of course been economic, cultural and social: a disruption to traditions, employment and family. The disruption is also to shared sense-making practices. Knowledge constituted in and along the trajectories of globalisation spills outside the institutions that might lend it authority or stabilise it.

Since the early 2000s, Oxford House in Bethnal Green in East London has been a key node in the networks of the Somali diaspora, performing civic and social functions within the UK, within Somalia and more broadly internationally. The practices of the various organisations and individuals working across these networks produce patterns of sense-making that circulate through a dynamic set of transnational social forms. Oxford House, a relatively small and intermittently resourced community and arts centre first set up by student missionaries to the impoverished East End in the late nineteenth century, has been globally influential in providing a base for Kayd – a Somali arts organisation – and the Somali Week Festival which has taken place annually since 2007. In turn, Kayd is closely connected with the Redsea Online Cultural Foundation and the Hargeysa International Book Fair. These organisations play an ongoing role in promoting Somali theatrical and poetic forms transnationally, both in Somali and in translation, as responses to a globalised and 'disrupted' experience.

Khat Out of Sight Out of Mind exemplified the way theatre and translation have been central to the emergence of sense-making practices at Oxford House. The project responded to concerns about use of the stimulant khat, a plant grown in East Africa, within the Somali community and possible links to health problems and other community concerns at a time when khat's legal status and associated pharmacological and social status constituted a particular problematic within the UK. The Advisory Council for the Misuse of Drugs (ACMD) were twice commissioned by the UK government to examine the health and social impacts of khat use and to make recommendations on its legality and produced reports in 2005 and 2013. Both times the ACMD advised against banning khat but made other recommendations in relation to social issues associated with its use (2005, 2013). Despite this expert advice though, in 2014 the UK government scheduled khat as a class C drug and as such its use and distribution were made illegal. The rejection of the expert recommendations by the government reflected, albeit in different form, the extent to which the issue also divided the Somali community within the UK. Many commentators attributed the government's decision to pressure from sections of the Somali community and desire to conform with the legal restrictions on khat already present in other EU countries as well as the USA. My argument here though is not really to do with the social impact of khat but rather to do with what arts practices might offer as a response to such an issue, particularly in their potential to assemble diasporic and globalised knowledge through theatrical approaches to performance and translation.

One of the challenges for the ACMD researchers and for the project at Oxford House was how to reconcile competing ways in which diverse individuals and groups might make sense of khat within different social, scientific or cultural frames. While the globalised experience of many people across a Somali diaspora can be summarised as in my brief account above, this account risks homogenising diverse experiences and eliding the ways that class, gender, race, generation or other factors may cut across Somali identity.[8] For the sociologist Avtar Brah, 'diaspora space' is where economics, politics and culture come together as historically dynamic processes within particular sites (1996, 181). Diaspora space both produces and is produced by particular historically and geographically contingent forms of knowledge and by groups and individuals who might or might not identify as Somali. *Khat Out of Sight Out of Mind* responded to these challenges of making sense within diaspora space both in how it assembled collectives and also in its creative practices.

The project was initiated and managed by Ayan Mahamoud (managing director of Kayd from 2007) as part of the Somali arts programme then running at Oh! Art funded by a small grant from Arts Council England. With project funding from the Wellcome Trust, Mahamoud assembled a group including Eleni Palazidou, at that time Consultant in General Adult Psychiatry at the Royal London Hospital, the poet and playwright Ali Raabbi Seenyo, director Zoe Svendsen, poet and translator Said Jama Hussein, the cast of five performers, as well as academic Martin Orwin, who chaired the post-show panel, and a video production company. Members of this group collaborated on the production of two versions of an original play performed in Somali and English respectively, a series of public and private discussions and video documentation. Both versions of the play told the story of a Somali man, Xirsi, living in London. It explored the impact of his khat use on himself and also on those around him, his immediate family and friends.

On one level the narrative seemed to be a polemic espousing the views and reflecting the approaches of those campaigners who advocated for the criminalisation of khat from the early 2000s up to the ban in 2014. Two of the characters in the play, Xiyaad and Xiriir, are anti-khat campaigners, who more than once express their frustration with the decision of the UK government not to ban khat after the 2005 ACMD report. The play graphically depicts social harms associated with excessive khat use through the actions of the two male khat users, Xirsi and his friend Xiddig, and in the rhetoric of the anti-khat campaigners. In one scene, Xirsi takes money which his wife has been entrusted with to pay for a funeral in order to buy khat. In another scene, anti-khat campaigner Xiyaad says she went to a mental health clinic and discovered that three-quarters of the people there were young Somali men. She also recounts a story of a man who inadvertently suffocated his three-week-old child while in a khat-induced stupor. In a scene involving Xiddig and Xiyaad, Xiddig is seen to suffer from an attack of paranoia brought on by excessive khat use. At the end of the play, Xirsi is dramatically, comically and somewhat incredibly reformed and redeemed when he decides to give up chewing khat on discovering that his children have told their school he is dead because they hardly ever see him.

Informal conversations with many of the creative team and statements made in the video documentary about their process suggested they were all in broad agreement about the harms of khat use to individuals and the community and in favour of a ban. Their positions and the many inferences in the play about a causal relationship between khat use and various social harms seem to run counter to the carefully researched and worded ACMD reports which found relatively 'poor' evidence for such a causal relationship (Advisory Council on the Misuse of Drugs 2013). The 2013 report summarises its position by stating that 'save for the issue of liver toxicity, although there may be a correlation or association between the use of khat and various negative social indicators, it is not possible to conclude that there is any causal link' (Advisory Council on the Misuse of Drugs 2013, 4).

However, I am not quite arguing that the project constituted anti-khat activism or a one-sided representation of a consensus that according to many commentators did not exist within the Somali community let alone more broadly (Thomas and Williams 2014). Neither am I suggesting that the play was straightforwardly didactic. As mentioned above, the project as a whole consisted of more than one element and even the performance event comprised more than one piece of theatre, more than one aesthetic framework and more than one set of theatrical conventions. The project commoned senses of khat by locating them within novel practices of linguistic and theatrical translation, navigating and innovating complex, sometimes contested social forms, bound up as they were with globalised experience and dispersed within diaspora space.

The Somali version, written and directed by poet and playwright Ali Raabbi Seenyo, drew on traditions of modern Somali theatre, as a set of forms that have historically made sense of experiences of change socially, aesthetically and lyrically. Afrax (2013) has argued that in the post-independence period, theatre in Somalia articulated transitions between traditional, pastoral, rural lifestyles and modernised urban living. This was manifested in the development of a national theatre, forms that mixed literary and oral approaches to theatre-making, the combination of different kinds of verse with prose and various modes of theatricality.[9] Seenyo works with and draws on the breadth of Somali poetic expression, *maanso*, with its range of different forms – including *gabay*, *jiifto*, *hees* – that are differentiated in their use of alliteration and prosody. In his play, characters spoke in verse interspersed with prose dialogue sequences. This hybrid form was also reflected in the staging. At times, performers faced the audience and spoke into microphones reciting the poetic narrative. At other times, they turned to each other and interacted physically through gesture or movement within a more mimetic, social realist mode.

The English version was translated by Said Jama Hussein and directed by Zoe Svendsen who admitted not to have understood a word of Somali. In this version, the prose improvisation predominates with much less use of microphones and more evidence of acting within the social realist mode. However, within the English version, instrumental music and songs in Somali by Jookhle, the performer who played Xirsi, remain, punctuating the action. A summary of the story by Jookhle told directly to the audience frames the piece at the start and in both versions the cast performed a song and dance at the end. In fact the formal relationship between the two versions is probably more complex than these descriptions allow in that theatrical features, blocking and other conventions seem to have bled between the two versions. As the same performers acted in both versions this was probably inevitable whether or not writer, director or translator might have conceived the staging in this way. The memorised verses in the Somali version are likely to have offered some kind of narrative structure to the English version even when they were not spoken in it. Blocking laid down in rehearsal of the English version reappeared in the staging of the Somali version.

Added to this, the performances of the actors in each of the two versions also reflect their own experiences of globalisation and inflect their playing of their roles. Jookhle, who played the main character Xirsi, was a relatively well-known actor and musician who had won performance awards in Somalia in the 1980s prior to the civil wars. As such his performance makes a connection with Somali performance culture in Somalia in the context of a period in which a national post-colonial theatre integrated traditional poetic traditions with the influence of other African and international theatre traditions. Unsurprisingly, Jookhle was a confident performer in both versions but clearly more at home singing and speaking in verse in Somali than improvising in English prose. The other performers, from a different generation, perhaps born or, more likely, educated predominantly in the UK, were fluent English speakers. The two much younger performers playing anti-khat campaigners Xiyaad and Xiriir were bilingual, speaking English with East London accents. One of them commented in the documentary: 'I'm more confident in the English one because it doesn't have any poetry.'

Ostensibly, the generational (and to a slightly lesser extent gendered) divide reflected in language use and in a relationship with a traditional poetic culture correlated with a political position on the effects of khat. One of the younger performers commented in the documentary that 'it's going to be tough to break that ideology, the fact that khat is not seen as a harmful stimulant, it is seen as part of our culture, part of our tradition.' This comment both stakes out a position but also acknowledges the multiple ways that sense might be made of khat within a diverse, globalised community and within diaspora space.

This collaborative sense-making contained disagreement and contradiction within its theatrical form of translation. At one stage of the performance, in both English and Somali versions, Xirsi's wife Xamdi challenges her husband about the missing £200 intended for his friend's funeral expenses. As she berates him, gesticulating with frustration, Jookhle as Xirsi, with impeccable comic timing, waits a beat, draws a breath and launches into a piece of music. He answers her verbalised concerns with a melodic and rhythmic response. Music communicates where words fail him. His mastery of the *hees* form of song sustains him, protects him, but also acts as a barrier to his wife's concerns. It represents but also sits alongside khat's psychoactive and psychosocial effects. Later in a post-show discussion Jookhle related how he continued chewing khat while acting in the play – he actually chewed khat on stage and in rehearsal. Therefore, to some extent, the performance event was a psychosocial effect of khat use, even though its effects within the plays were exaggerated and theatricalised by the actors playing Xirsi and Xiddig. Jookhle commented that while he knew it was bad for him and that being in the play had encouraged him to think about stopping, he was unsure how to. His khat use on stage, both represented and presented, underpinned his performance. Taking it away is a complex, destabilising and unpredictable proposition that is, of course, rehearsed at the end of both plays.

The political potential I see within this project's practices lay not so much in what it seemed to have to say about khat's social or pharmacological effects. These facts still seemed to be disputed or unavailable. Rather that its capacious social and aesthetic form in which translations sat beside and permeated each other was able to hold together knowledge about khat located within different life experiences and different knowledge traditions, including but not limited to: psychiatry, pharmacology, sociology and the complex prose and poetic forms of contemporary Somali performance. Even if the narrative seemed to run counter to the ACMD reports in certain respects, rather than opposing it with alternative facts, the project explored the epistemological gaps created both by contradictions between different understandings of khat's effects, the absence of knowledge or indeed the apparent dislocations of existing knowledge from contextualised life experiences.

Khat Out of Sight Out of Mind made sense of khat in terms that were importantly different from the ACMD report, although in its themes and performance aesthetic it also reflected some of the report's conclusions. The covering letter of the 2013 ACMD report commented that:

> It is apparent from the evidence on societal harms that it is often difficult to disentangle whether khat is the source of community problems or, to some extent, its prevalence and use is symptomatic of the problems for some individuals and groups within the community.

The project contributed to an ongoing, evolving tradition in which lyrical forms of performance make sense of complex experiences of globalisation. Disentangling khat, or rather coming to terms with its entanglement, might find an untapped resource in poetic sense-making that actually has a rich tradition not just reflected in *Khat Out of Sight Out of Mind* but in other works such as Cali Ibraahim Iidle's (1968) play *Dhulkeenna Dhibaha ka Jooga* (*The Troubles Present in our Land*) to Asha Lul Mohamud Yusuf's (2012) poem *Gocasho* (*Recollection*).[10] *Genesis* and *Goodstock* translated dominant narratives about hereditary cancers and their prevention and treatment by mapping them onto and across people's lives, families and relationships. *Khat Out of Sight Out of Mind* showed how awareness of and experimentation with diverse performance traditions enabled the translation of a complex issue across diaspora space. In all these projects translation also coped with tensions between the senses shared within and across different groups as determined by ways of knowing and the time-spaces that shape identity.

Colour coded

And so I return to the grime MCs and dermatologists in Birmingham. Breast cancer knowledge has been translated both by molecularisation and by the different ways it has been problematized by clinicians, scientists and those who have experienced it as patients or carers. Knowledge about khat and its psychoactive effects has been dispersed, dislocated and disturbed by war, coerced migration and the challenges of linguistic and cultural translation. Knowledge about the skin has similar, albeit more protracted, stories of dispersal along power-geometries of time-space. Cultural knowledge about skin, often given form in the discourse of race, remains ambivalently connected or detached from scientific, and therefore increasingly molecular, knowledge about it.

The translational challenge faced initially by Matthew Walters in assembling the MCs and the dermatologists was discursive and spatial. It was a practice of assembling disparate groups with disparate expertise in relation to a shared object of interest within shared social space. This space was roughly delineated by the urban environment of Birmingham with its internal boundaries and routes across it. However, assembling young people from Handsworth and dermatologists in the same social space is by no means straightforward. In Walters' case it relied on mobilising a complex social network with its own historical particularities. Central to this network were a number of creative organisations that were only accessible to a relatively young creative producer from Handsworth, by virtue of transient, shifting funding regimes and the skills, ingenuity and creativity of Walters and those who supported him. At this time, there was a growth in political support for creative and cultural education for young people both within and outside schools through project funding and other support from government arts and education initiatives often via and in partnership with independent charities. In the early 2000s, the activities of government-funded organisations like Creative Partnerships (2002–11) and First Light (2001–13) coincided with and often complemented funding and support for public engagement with science initiatives.

In addition to the young people directly involved in the various creative outputs of the project and the dermatology team, there were key facilitators including the music producers De-cifer Sounds, independent filmmaker Justin Edgar, the film and TV production company Resource Base, the arts venue the Drum, Sutton College and Don Kinch of Nu Century Arts. This infrastructure was assembled just in relation to this project and I briefly list it here to show the extent to which new modes of knowledge production that attempt to address cognitive injustices require significant and sometimes circuitous social creativity to navigate around access barriers. It was not as simple as taking the bus from Handsworth.

A notable feature of this project, shared by other comparable projects, was also the way it employed and proliferated social, media and aesthetic forms of expression. Divergent and eclectic forms of expression responded both to disparate ways of making sense of the theme and explored different modes of engagement with potentially diverse audiences. Walters and his collaborators produced a DVD that brought together the various elements that emerged within the project and presented them in parallel rather than as an iterated final performance. The DVD is a coherent creative output that deals with skin colour and how it is made sense of by the young people and dermatologists. However, it contains expression in multiple idioms presented as a collection of short films. There is *Struggles*, a short music video with lyrics and performance by the grime MCs known at the time as Merky Movements put together with Walters, Edgar and the music collective De-cifer Sounds. Young people from Sutton College made a short video drama called *Fake*. Documentary features included: *Views*, a series of statements by different people in response to simple questions about skin colour; a filmed conversation between young people and a dermatologist; and another informal conversation among a small group of young people.

The DVD as medium of performance documentation enabled the multiple idioms used to explore skin colour to be located across various parts of the city. Merky Movements rapped on the roof of a red-brick factory or warehouse with views across the post-industrial cityscape. Shots of the MCs are intercut with a sample of Birmingham's skin tones represented by shoppers in the city centre. *Fake* plays out within a further education college and in the homes of the young people. The dermatologists are clearly on home turf when meeting the MCs in an unmistakably clinical office. An intimate and engaging short conversation takes place between young people in someone's back garden. *Views* is removed from identifiable social or physical location by a green-screen technique and located instead on a colour spectrum: colour coded as in the title of the DVD or chroma-keyed in the film-making idiom.

Colour Coded offered an approach to translation that did not attempt a single, simple linguistic substitution or spatial relocation. No one location or idiom is privileged in relation to another. Skin colour was mapped onto multiple idioms and multiple locations on a DVD and in a city. It resisted the attempt to make dermatology and grime completely commensurable as ways of making sense of skin colour, although it did attempt to co-locate them both within Walters' social infrastructure and in the various creative practices. The different idioms make sense of skin colour in ways that overlap; together they articulate tensions but also provide a rich resource. *Struggles* interweaves three different strands: shots of shoppers posing in Birmingham city centre; a narrative showing ways in which young black men feel determined as criminals, in other people's eyes, by their skin colour; and the voices of the MCs who offer an alternative lyrical account rejecting narratives of epidermal determinism and asserting an alternative relationship with their skin. *Fake* tells the story of two young women who respond in different ways to the pressures of hegemonic ideas of beauty that attach particular value to lighter skin tones. Such ideas permeate marketing for cosmetic products, some of which can be harmful.

The *Struggles* music video reflects both the challenge of translation but also how different idioms or codes might resource sense-making. A key feature of grime as a way of making sense of experience is the way that MCs locate themselves as members of collectives and within a particular city or area. What they have to say – their lyrical idiom – is heard in specific relation to their location in time and space. Grime is about making yourself heard within a local neighbourhood and about locating the neighbourhood within a broader geography. Grime has a fairly precisely located history in particular areas of the UK's larger cities. Most commentators place its emergence in parts of Bow, East London, specifically in the work of Wiley, but it quickly spread to other parts of London and other major cities in the UK. Its development has been important in Birmingham, where several high profile artists, notably Lady Leshurr and Jaykae, developed successful careers from the early 2000s. *Struggles* announces itself, locating its lyrics in this way: 'Yeah, De-cifer Sounds,

Merky Movements… Struggles, yeah… coming straight from Birmingham.' Grime MCs articulate specific aspects of urban life for young people living in low-income, multi-ethnic estates negotiating linguistically between local experience and a relationship with the wider world (Adams 2018). *Struggles* locates its content, themes and information as belonging to De-cifer Sounds and Merky Movements and coming from Birmingham. Even the identity of the group 'Merky Movements' locates them fairly specifically within grime as specific globalised and localised sub-culture. The word 'merky' has been traced back to US hip-hop in the 1990s and its meaning, or rather translation, has shifted from 'killing' or 'dominating' to the artist Stormzy's use of it in 2018 to mean 'excellent'. In Birmingham in 2006, I think the MCs were using it more in this latter sense although the title of the track certainly links the desire to excel with the need to struggle ('Weekly Word Watch: Avozilla, Hooyah!, And #Merky' 2018). In the context of the project though, the name of the group also resonates, through the homophone 'murky' with a concern with the relation between colour, movement and struggling to find clarity.

The project articulated location, co-location and dislocation as translation practices in various ways alongside its frequent linguistic creativity. In a filmed question and answer sequence with dermatologist Celia Moss, questions about acne, albinism and skin colour were asked by young people in one location and then answered in another – the professor's office – questions and answers are co-located in post-production. In the meeting between the MCs and the dermatology team, they sit slightly awkwardly around a table in what looks like a hospital meeting room. 'How come my skin's black and others are white?' asked one of the MCs. Scientific answers to such questions appear to have no location outside the institutions of learning or research dissemination. This process is governed by genetics in complex ways that are not fully understood. Such facts and their limitations are clearly central to the basic education of dermatologists but, in this idiom and within the spaces of dermatology (the clinic or academic journals), they appear to have limited relevance to everyday life. Gregory Barsh has written in the pages of *PloS Biology*, 'skin reflectance is determined by the amount and distribution of pigment granules within keratinocytes' (2003, 20) and in *Trends in Genetics* a few years later Richard Sturm explained how 'SLC24A5 promotes melanin deposition through maturation of the melanosome' (2006, 464). But, aware as he might have been of the work of Sturm or Barsh, the member of the dermatology team does not say this. He replies: 'it's the genes that we have that make us the colour we are.' He relocates the phenomenon from pigment in keratinocytes to both the inclusive, organic and humanistic scale of the first person plural at the same time as retaining an orientation towards the molecular causes of this identity characteristic. In a hybrid idiom he relocates the subject as 'we' who become active possessors of genes and who 'are' a colour. Knowledge produces being. Objects make subjects. Epistemology shades into ontology. While skin colour is a matter of significant concern to the MCs from Handsworth and the dermatologists' matters of fact seem to be of interest, there is a significant gap between the location of the fact within clinical practice and institutions and its location within the lives of young people.

The music promo and other modes of performance in *Colour Coded* experimented with this kind of play between subjects and objects, ontology and epistemology in various ways. In *Struggles* there are a number of shots in which shoppers, apparently stopped in the busy city centre, pose alone or in groups. They hold up signs against their bodies: 'I've got lots', 'I've got some', 'I haven't got much' (Figures 5.2a and 5.2b). These are cryptically coded messages. Shot in front of a designer shop in a relatively upmarket part of Birmingham's shopping district, is this a socio-economic valuation? When read against their skin, in the context of the project, they seem to refer to melanin or Barsh's 'pigment granules within keratinocytes'. But it could be either and the coding perhaps begs questions about the relationship between the two quantitative frameworks of economics and pigmentation that are often used in various kinds of qualitative judgement.

Dermatological codes for skin tone are juxtaposed with grime's own codes for skin. The term grime itself advertises its concern with what covers the surface and its fearlessness in addressing unsanitised aspects of experience and difficult, messy feelings. Grime has sometimes been defined by a move away from mainstream hip-hop's glorification of bling lifestyle, but it cannot just be seen as a local urban British translation of globalised African-American culture (Bramwell 2015). It is a form that values and validates the felt experience of diverse ordinary young people and translates a variety of globalised and localised ways of knowing, making sense of everyday experience, commenting on this and articulating ambitions in direct relationship to this experience.

When quantitative statements about skin colour are held against bodies, a mode of molecular visualisation – difference at a molecular level – is juxtaposed with vocalised narratives about the felt experience of skin colour within a social context that is both contradictory and also in subjective tension with the apparently objective statements in the visuals. In the first verse of *Struggles*, one of the MCs' raps:

> I'm proud to be black,
> If you ask me it is the greatest feeling,

Figure 5.2 Screenshots from *Struggles*

> I don't like getting victimised,
> I get watched in shops cos they think I'm stealing,
> I'm not stealing, I'm a proud black boy and yeah my status is appealing,
> You're never gonna bring me down,
> My head's up there always higher than the ceiling.

But it is not the apparently objectifying dermatological statements here that result in the victimisation. Everyone's melanin is objectified – difference here is just a question of quantity or proportion. Sixteen-bar solo rapped verses are punctuated by the collective chorus 'Living with the struggles, life is hard, life is tough and you know we're reppin double'. Repeated rhymes 'feeling… stealing… appealing' and repeated rhythms and scansion are held together by a 140 beats per minute backing track.

This is grime's poetic code. The code though is located in relation to everyday experience and aspirations in its content and via the settings of the video. The second soloist raps behind a desk, pencils in pots and others writing: 'I'm with Merky Movements, I wanna be a big, big star, I wanna make big, big bars'. This desire for fame and creative fulfilment is set in a school or college then against the shopping street and a woman with a sign, back to the school and then onto an urban roof-top. The rapper locates himself within a poetic code, an education system, within his crew, within the city and in relation to dermatological code. As the verse becomes the chorus, the solo rapper merges back into the crew vocally and in the next shot.

In *Struggles*, the resistant, aspirational ethos of grime and hip-hop more broadly remains, but in place of the way of denigrating another's skin sometimes resources rap battles between MCs, so rehearsing the skin as a site of value judgement, here actual skin tone becomes a matter of quantity (of melanin) rather than quality. At the same time the track does not avoid the facts of racism. The MC's experience of being 'racially profiled' in shops is an inescapable subjectively experienced social context for dermatology's objective measurements.

Struggles gives a glimpse of how grime and dermatology might combine translations, relocations and code-switching to find an idiom to resist or at least reframe racism. In linguistic work on code-switching, different language systems are juxtaposed in spoken interactions normally within multilingual contexts (Gumperz 1982; Kolehmainen and Skaffari 2015). Code-switching can sometimes be intended to reiterate or emphasise meaning (Harjunpää and Mäkilähde 2015). Translation, on the other hand, often implies transformation: the movement of meaning from one code to another (Kolehmainen and Skaffari 2015). Here there are a series of translations from the dermatologist's explanation of skin genetics, what it means to have black or brown skin, to the inscriptions on the signs held up in the street and the MCs' articulation of what it means to be black. *Struggles* though and the Colour Coded project more broadly also switches back and forth in a rhythmic movement between these different codes rather than the linear uni-directional movement suggested by some conceptions of translation, arguably including Callon's.

Colour Coded locates its translations and code-switching in emergent social forms. The social form of the project both contains and is contained by its spatial and linguistic codes. Grime's aesthetics – the content of the lyrics and visual imagery used in videos – is inextricably linked with its social forms. From its outset, on pirate radio stations at the top of East London tower blocks or outside takeaways, unknown young artists put together their own crews and disseminated their tracks themselves through various modes of file sharing and social media as they became available. Notable figures such as Dizzee Rascal and Stormzy have achieved mainstream success and become high profile figures within the music industry. However, a certain counter-cultural ethos still exists within even the work of the biggest names. Wiley shared his entire back catalogue and sacked his record company exemplifying what Bramwell has called grime's 'ethics of free exchange' (2015, 30). Walters and Merky Movements were drawing on these socially and artistically entrepreneurial features of the genre in its relatively early days in taking on complex questions about skin and placing what they knew in direct relation to authorised accounts of skin from science and society. A key resource of grime as a poetic genre and a social form is its lexical ingenuity and flexibility. Words such as 'merky' appear, move and change meaning in subtle but potentially rapid ways over time and across space.

In the DVD there is no evidence of particularly extensive reciprocal exchange between different groups. The filmed conversation between the MCs and the dermatologists is short and restricted to a few questions from the MCs and some brief answers from one doctor. However, when I got back in touch with Walters and Moss 12 years after the project when writing this chapter, both attested to the significance and legacy of this project for them as individuals and more broadly. It had an impact on the disparate representational practices of research, clinical practice and filmmaking, however hard this may be to pin down over a decade later. In some ways the claims I am making for this kind of practice may seem limited. However, the continued unlikeliness of finding dermatologists and grime MCs on the same bill suggests that translating skin colour still has a long way to go. As I mentioned at the start of this section, science has a highly ambivalent relationship to discourse on race. A strand of work in contemporary genetics research confidently asserts that there is no biological basis for race as a concept, yet this argument is almost drowned out by the continuing legacies of centuries-old scientific support for colonial racisms, popular web-based companies offering racial profiles on the basis of a self-administered DNA test and a number of other ways in which race and science are re-imbricated (Thompson 2006). Making sense of skin colour needs the lexical ingenuity of grime as much as it needs the quantitative analysis of dermatology and it needs translation to take place between such colour coding.

This is not about setting knowledges up in opposition, to turn again to the work of Boaventura de Sousa Santos who has argued that 'finding credibility for non-scientific knowledges does not entail discrediting scientific knowledge. It implies, rather, using it in a broader context of dialogues with other knowledges' (Santos 2014, 193). Translation, however, needs to go beyond dialogical forms to produce time-spaces that cut across boundaries variously delimited and enable the forging of solidarities and even the possibilities for contestation over the objects of knowledge. If Theatre of Debate innovates a process for enabling the politics of life to be played out within institutional space, then in this chapter, the space itself becomes part of the aesthetic. Just as ways of life are generative of form, here so form is seen as generative of ways of life. Commoning sense demands recognition that people do not just acquire or gain knowledge, but live it.

Notes

1 The historian of science Joseph Needham (2000) made a major contribution to debunking the narrative in which the origins of modern science are seen as exclusively Western in origin. There are many examples of how biomedicine has retrospectively validated and sometimes expropriated traditional medicines from different parts of the world (Leach *et al.* 2005).

2 There is a range of scholarly work that explores medical and cultural histories of cancer. Jackie Stacey's (1997) *Teratologies: A Cultural Study of Cancer* intertwines her own personal narrative with critical commentary on biomedical and cultural representations of the disease. Work by Susan Lederer (2007) has discussed cancer narratives in Hollywood films and examined how this influenced popular perceptions of the disease and knowledge about it. Siddartha Mukherjee's (2010) *The Emperor of All Maladies: A Biography of Cancer* weaves together narratives from different perspectives as clinician, biologist and historian.

3 One of the most famous human cell lines used in thousands of scientific research studies has been that originally taken from Henrietta Lacks, who died of cancer in 1951. Her story, and that of the 'Hela' cells derived from her body, offers a significant perspective on a number of complex social, ethical and political issues that are explored in an important book by Rebecca Skloot (2011). Lacks' story has been adapted for the stage in the USA in a 2018 play by J. Nicole Brooks (Reid 2018) and in the UK in a 2012 solo performance by Adura Onashile (Ranscombe 2013). Other artistic responses to the history are discussed in an article by Adele Senior (2011).

4 Particularly influential examples in the UK might include in the 1970s *Through the Night*, a BBC TV drama work by Trevor Griffiths (1977), Louise Page's 1978 play *Tissue* (1990), in the 1980s the photography of Jo Spence, and in the 1990s work by artists like Sam Taylor-Wood (Bolaki 2011; Toon 2014). In the USA, notable examples might include Margaret Edson's 1995 play *Wit* and the artist Matuschka's 1993 cover for the *New York Times* magazine. More recently work by Brian Lobel (2008) solo and with Bryony Kimmings (2018) in the UK has produced particularly challenging and critical translations of cancer. There have also been numerous soap opera storylines involving cancer and its treatment.

5 A lot of literature about genetic breast and ovarian cancer discusses the peculiar ontological status of the person who has been tested and found to have a genetic predisposition. The term 'previvor' is widely used instead of patient to account for this peculiarity, which in excellent articles Coleman Nye (2012) identifies as 'theatrical' and Diane Herndl as 'virtual' (2014).

6 This could be seen as a reworking of the theme within many science plays of the impossibility of unknowing that is explored within canonical productions like *Copenhagen* (Shepherd-Barr 2006).

7 Hirst reflected on her experience in a piece in the *Guardian* newspaper (Hirst 2015).

8 Thomas and Williams (2014) suggest that part of the problem with evidence on the harms of khat use emerge from the 'super-diversity' of khat users.

9 Somali performance culture in diaspora space is almost entirely neglected by scholarship in theatre and performance although there have been some higher profile examples of practice in the UK since the production I discuss here, e.g. National Theatre Scotland and Contact's production of Yusra Warsama's *Rites* (2015) and National Theatre Wales's *De Gabay*, produced in 2013.

10 Iidle's play has been discussed by the leading scholar of Somali poetic forms B.W. Andrzejewski (2011).

6

Performing cosmopolitics

Most people have some kind of experience of motion sickness from travelling as a child in a stuffy car, staying on a swing or roundabout too long, being in a boat on a rough sea or on a turbulent flight. It grows as a feeling of nausea, perhaps with a slight headache and clammy skin and leads to feelings of weakness, drowsiness or apathy. It is perhaps slightly puzzling that people do not usually develop motion sickness from walking or running. While it has a long history, the possibilities of motion sickness seem to have proliferated alongside technological developments. Despite this long history and the deleterious effects of motion sickness on militaries, economies and nice days out, the feeling has not been eradicated either through pharmacology, engineering or lifestyle. Prolonged exposure to the same motion can mean you become habituated through what sailors call getting your sea-legs. However, the only sure fire way to avoid seasickness remains to stay on land.

This book has attempted to explore how performance, and forms of artistic or theatrical performance in particular, may resource negotiations with the sometimes sickening upheavals of scientific and technological development that I introduced in Chapter 1. It has taken on a broad range of scientific and performative practices, because I would argue that there is a need to find diverse, creative ways of practising cosmopolitics. Doing cosmopolitics, for me, means facing the challenge of coming to terms with different ways of knowing. A lot of problems faced in the world are cosmopolitical problems: they are problems that cannot be solved by singular ways of knowing, neither indeed by the application of thought, feeling or action separately. Science, technology and public feelings are deeply entangled at various levels. While climate science and breast cancer may occupy different faculties of a university, and the businesses of making music or manufacturing skin-lightening cream operate in different sectors of the economy, such practices can easily come together in the lives of one person or family. Politics in people's lives as well as politics at a global level are often, or at least should be, cosmopolitics.

But the demands of the cosmopolitical often leave us feeling at sea. Developing more equitable cosmopolitical practices might often mean upsetting or crossing disciplinary boundaries, building bridges between institutions and other collectives, or searching for approaches to translation between discourses or disciplines. Such practices are creative and imaginative. They require the development of processes of co-production or, as I have reformulated in the previous chapter, processes for commoning sense. I have looked for these processes in the theatre and related artistic performances, because I have argued that theatrical conventions can help make common sense of the experiences of performance pressure and of science in performance. The history of theatre is a history of negotiations between reality as it might be perceived within and without certain spatial and temporal conventions. I therefore think its practitioners might have something to offer such negotiations. I have also written this book because of a feeling I had about some of the practices I have documented here. This was a feeling that these practices and the practitioners collectively involved had found something out in the course of their work. However, what they had found out was not necessarily clearly recognised or valued. Many of the practices discussed in the chapters of this book have been rather neglected either as artistic practices or as practices of science communication. I suggest this is partly *because* of their value as cosmopolitical performance. Simply put, because they might be considered art, science, technology, education *and* communication, there is a tendency within disciplined research to undervalue them as art, science, technology, education *or* communication. If it is important for grime MCs and dermatologists to have conversations about skin, as I believe it is, there may be a need for some undisciplined research as well as practice.

Chapter 3 attempts to bring into cosmopolitics practices that might already be valued commercially or artistically but that I attempt to revalue as ways of knowing. My conclusion is not that theatrical processes for commoning sense are necessarily to be found at the margins or among the avant-garde, they may just as well be within popular culture. In some cases, as in my discussion of the vertigo aesthetics of Headlong Theatre, commoning sense occurs through a kind of diagnostic process. As a collective practice of speculation, theatre can help a reflection on the feelings provoked by an awareness of climate science or neuropharmacology. Theatre, even Broadway musicals, can also ask provocative questions about why particular forms of knowledge may cause such mixed up feelings. At the same time, theatrical performance can constitute particular forms of intervention into the politics of knowledge. This may be through enabling particular publics to take science seriously enough to laugh at it, as Gob Squad seemed to have achieved through their complex partnerships with roboticists and opera singers.

Some scientific accounts of motion sickness refer to the way the body becomes overwhelmed by stimuli coming from different senses – vestibular, proprioceptive and visual – and account for it in what is called 'sensory conflict theory' (Dobie 2019). Seasickness can be precipitated or exacerbated by anxieties and also by memories of prior experience and indeed by imagination and anticipation. Getting your sea-legs can reflect the body acclimatising itself to such conflicting sensory experience. At the same time, removing or modifying some of these confounding stimuli, perhaps by getting some fresh air or locating the horizon, may also be necessary to stop the nausea. Chapters 4 and 5 have discussed ways that artists can facilitate processes of making sense of complex and sometimes contradictory knowledge perhaps rooted in apparently incommensurable knowledge systems. Such process may involve the development of somatic expertise and also skills in translating different ways of knowing across space and between languages or cultures.

I want to close this book with a brief discussion of the work of a couple of artists who have taken on motion sickness both literally and in the metaphorical terms that I have been gesturing towards in my discussion above and indeed in the idea of vertigo aesthetics. They both have also taken on climate change, perhaps the cosmopolitical problem of our time, producing outcomes that common sense in similar ways to those I have already discussed. The US singer-songwriter Cynthia Hopkins and the UK choreographer Siobhan Davies work in very different media. However, both have had an extraordinary experience, facilitated by the charity and cultural organisation Cape Farewell, having travelled on sea voyages to the Arctic in a fairly unusual kind of artist residency. I speculate here that there may be features of such a residency process that hold broader cosmopolitical potential.

It is not a new idea for artists to draw on the Arctic for inspiration, either through their own or more often others' travel experiences. It frequently appears as an imaginary within Victorian literature, painting and popular culture, particularly around the time of the disastrous loss of the 1845 Franklin expedition. Scholars of this period have examined work, including paintings and literature, via the aesthetic category of the Arctic Sublime. This category refers to the experience of awe and wonder at the environment. It reminds humans of their relative insignificance, but at the same time opens up a safe distance between (or within) the aesthetic experience and the experience itself (Loomis 1977). Scholars have critically explored how such artists, through the Arctic Sublime, resourced constructions of nation and gender that privileged masculine and imperialist accounts, thereby marginalising or entirely erasing the experience of others, notably the indigenous people of the region (Morgan 2016).

Such an aesthetic category is still ubiquitous in performative or theatrical representations of the Polar Regions, not least in the nature documentary form through which most people are able to access them. Political critique of the various ways in which the Arctic may be depicted as sublime is therefore still of great value. In fact, both the examples of performance practices I am going to discuss here respond in some way to such traditions and could well be analysed in such terms. However, I am more interested in the way that artistic practices and indeed their critique might go beyond such patterns. How can cultural practices re-examine the safe distance implied by the Arctic Sublime, when climate science suggests there is no safe distance from climate change? What is the relationship between the delusions of the sublime and disciplinary knowledges? How might practices explore a variety of possible human relationships with, and even agency in relation to, the environment and the nonhuman more broadly? In what ways might artistic and critical processes emerge as Benjamin Morgan (2016) puts it 'after the Arctic Sublime', not necessarily temporally 'after' but as another, additional possible perspective?

Cape Farewell has reflected a particularly systematic initiative to use ship-based residencies of artists, scientists and others to generate artistic and scientific representations of the Arctic but also to stimulate a range of public responses to these representations. As such, it is an experiment in creating a cosmopolitical process. In some ways Cape Farewell seems to rehearse the Victorian voyages of adventure with their diverse and sometimes spectacular cultural outcomes. It is clearly an ambitious, costly and even hubristic enterprise that potentially lays itself open to charges of hypocrisy when considering its carbon, environmental or financial costs in relation to its apparently eco-political objectives. However, its relatively low-key DIY infrastructure – principally an old Dutch sailing ship – and its particular brand of imaginative interdisciplinary processes also seems to gesture creatively beyond the voyage as necessarily a voyage of conquest and its artistic outcomes also often seem to move beyond the sublime.

The Cape Farewell project was initiated in 2001 as a fairly specific cultural response by the artist David Buckland to version 3 of the Hadley Centre climate prediction model (HadCM3). Use of this model along with a vast array of other kinds of research led to conclusions by researchers about the contribution of 'anthropogenic forcing' to global warming featured in the successive assessment reports – particularly 3 and 4 – of the Intergovernmental Panel on Climate Change (IPCC) (Stott *et al.* 2000). One of the difficulties faced by climate scientists has been to cope with planetary scale when trying to understand climate change. Planetary scale of course spans space and time and modelling it involves combining large quantities of data from different sources. The HadCM3 model coupled atmospheric and oceanic components (Met Office 2001, 3). Later models such as the UK's first Earth Systems Model (UKESM1) add many additional components (Williams 2019).

Buckland *et al.* (2006, 6) have explained the invitation that was issued to artists and others in these terms, which refer to the planetary scale of climate science as a problem that I would suggest is a problem for cosmopolitics:

> The Cape Farewell invitation was open: come to the Artic, engage with scientists and, we hope, be inspired to make art. Art is never a guarantee; many a time I have followed a path of enquiry only to abandon it in frustration. What is extraordinary is that each artist has produced work that celebrates authorship. When unpacked, each artist has in some form responded to this cold, Arctic place and the way it is changing as the Earth warms. They have all told the story on a human, rather than a planetary, scale.

Buckland *et al.* suggest that IPCC reports and other scientific attempts to tell a story on a planetary scale are limited in their capacity to engage certain publics. His project wagers on the capacity of artists to respond to interactions with the environment and scientists who research it. He clearly hopes that their storytelling 'on a human, rather than a planetary scale' will then create wider public engagement with the findings of climate science and constitute a particular form of environmental politics conducive to social and ecological change. He sets sail for the Arctic, as it is there that the impacts of climate change are most readily apparent or maybe in search of a time-space where relations between human and nonhuman might be reconfigured.

Buckland's logic here is translational in spatial and discursive terms. His practice as a producer of translation is therefore analogous to the practices of Matthew Walters or Ayan Mahamoud discussed in the previous chapter, or Nigel Townsend's work on Theatre of Debate discussed in Chapter 4. However, as I hope is clear by now, I have not been arguing for artistic performances to be valued not just as communication or dissemination of existing knowledge, but rather as forms of knowledge co-production. I therefore want to explore how Cynthia Hopkins and Siobhan Davies might be seen as co-producing knowledge about living with climate change, contributing components to models, rather than simply communicating or aestheticizing climate science. I think their work is particularly interesting in the way the artwork they produced in response to their experiences of Cape Farewell manifested approaches to co-authorship. They tell stories and indeed conjure imaginaries that reach out towards the planetary scale as well as that of the individual human. They explore ways of assembling different collectives of humans and nonhumans that register sensory conflicts within their rather different sets of artistic conventions.

Sensory conflicts in the Cape Farewell process arise not just from relationships with the environment but also from the ways artists and scientists have tested each other creatively. These tests are both physical and intellectual. On one Cape Farewell voyage, the artist Dan Harvey challenged oceanographer Simon Boxall to use his understanding of tides and currents to predict the location of a block of plaster hung under a buoy so that the flow of water within the fjord would sculpt it. At the same time Boxall's attempt to measure the sea temperature and therefore to detect the Gulf Stream, meant stopping every 50 minutes in the rough sea: 'This was the point at which the scientist's popularity with the artists was put to the test as we pushed out into the heavy swell of the Arctic Sea' (Boxall 2006, 27). The discomfort of interdisciplinary collaboration here was visceral. The translation of data into sensory experience and back again was repeated and iterative. Such processes, for me, hold potential for exploring relationships between different ways of knowing and for registering the incompleteness of knowledge, not just in terms of incomplete data but incomplete awareness of what is not known. Being destabilised by a rocky boat is one rather direct way of experiencing sensory conflict and posing it as creative challenge.

Siobhan Davies has been a pioneering choreographer of contemporary dance since the 1980s. Since opening new studios in 2006, her company has worked as an 'investigative' arts organisation that supports a wide range of choreographic projects that go beyond touring dance productions (Siobhan Davies Dance n.d.). Davies was invited onto the third Cape Farewell voyage in 2005 which visited Svalbard. While on the expedition she left the ship and walked across the snow and ice, burdened down with layers of clothing and numbed by the cold. Sometimes she took other members of the expedition with her in a 'walking dance'. In these conditions of sensory conflict, dance became limited to the act of walking; even simple physical movements were intensified. The overall sensory experience for her was one of hyper-vulnerability. The need for protection or taking care became 'momentous' for her, so grave, weighty and of historical significance:

> The beauty, sheer physical beauty, is something that has imprinted itself upon me. But it is a skin over something far more ferocious. I sense a cold, I sense a vulnerability. I feel myself as something hot and bloody. My body, if it were harmed, the flesh would bleed. So I find the little bit of warmth I have, I need to protect it. The idea of protection, of care, seems particularly momentous here. (Davies 2006, 88)

The Arctic environment limited the range of physical expression she was used to as a dancer and choreographer and yet the feeling of being in a living but vulnerable body was intensified.

She played with this sensory conflict in a video installation she created for a 2007 exhibition at the Natural History Museum in London. She choreographed a short piece for the dancer Sarah Warsop entitled *Endangered Species*. In the piece, Warsop wears a costume designed by Jonathan Saunders that incorporates a number of flexible rods sticking out about an arm's length in several directions from her body. These rods both extend her movements beyond her body's normal outline, but also limit them. In the video installation, designed by Sam Collins with video by the filmmaker Deborah May, the dancer appears as a tiny hologram within a museum display case, seen through the glass within a polished wooden frame. The hologram of the dancer gradually becomes harder and harder to see.

The piece figures the individual human as an endangered species like others displayed at the Natural History Museum. On one level this is a transparent piece. Visitors walk round an institution redolent of Victorian confidence born out of science and empire. Davies and her collaborators place humanity alongside the other life forms humans have discovered, researched and sometimes destroyed. This addition to the collection though is barely there. It is a shimmering projection which struggles against its environment and then disappears. In its process it clearly does render the environment of the Arctic and understanding about the planet 'on a human scale'. However, the human here is figured as flesh and blood, a speck of warmth in a cold place. The human is not transcendent idea but endangered species. Doug Risner has theorised the production of knowledge through dance in the rehearsal studio as 'a spiralling process of self, society, body, memory' (2000, 169). By venturing out of the rehearsal studio into the Arctic and then into the museum, Davies expands this process to include a relationship with the environment and indeed the planet. The piece incorporates the sensation of being cold, being vulnerable, being in a vast inhospitable place. It also holds together an awareness of history and a relationship with the planet through an awkward reflexive shift where the viewer stands outside this immediate experience viewing this vulnerability and its complex histories shrunk to fit into a glass case.

Davies' work grapples with the experience of being part of a human species whose term of existence on the planet is finite. It plays with scale, through a multimedia aesthetic that speeds up a speculative modelling of human extinction, and through its capacity to shrink the human body rendering it almost invisible. It also grapples with 'species-being', in the sense used by Marx, that is both the sense of being a living species like any other, but one defined by its consciousness of being human as distinct from any other species (Wartenberg 1982). The viewer is positioned as a subject outside the history contained in the glass box, but perhaps aware of the historical structures – the box, the museum, the artists – that enclosed the body in the box. The speculative theatricality of the piece offers an experience of an imagined future and gestures backwards towards the museum's role in the co-production of an imperial natural history.

The piece perhaps even goes as far as attempting to address the challenge presented by the historian Dipesh Chakrabarty to mediate an 'experience [of] ourselves as a geophysical force' (2012, 12). That is to say, there is an extent when watching the disappearance of the human dancer, we are aware of making her disappear; it is almost like we get a glimpse of our own disappearance. The piece is elegant though and the joins between these perspectives are barely visible. Davies' process of coming to terms with these different modes of existence is hard to make out. In translating a particular sensory experience seamlessly into multimedia installation, perhaps she glosses over some of the sensory conflict. Chakrabarty argues for 'the necessity of thinking disjunctively about the human, through moves that in their simultaneity appear contradictory' (2012, 2). This means not only thinking on a multiplicity of scales, from the planetary all the way down. It also means thinking about different modes of existence and different kinds of (e)motion and ways of bringing these contradictions onto one plane.

The singer, writer and performer Cynthia Hopkins responded to her Cape Farewell experience in a rather different way from Siobhan Davies. Rather than attempting to find one imaginative process to manifest embodied experience, she developed a polyphonic performance piece that proliferates voices, narratives and forms of sensory experience. *This Clement World* was presented first at St Ann's Warehouse in Brooklyn in 2013. It was based at once on Hopkins' own visceral experience of being on the voyage, as well as documenting and developing a dialogue between a diverse cast of other characters. *This Clement World* combined documentary footage, filmed by Hopkins on her three-week trip to the Arctic in September 2010 on the Dutch schooner, the Noorderlicht, with songs and monologues by a series of characters all played by herself. The characters include real and fantastical figures who reflect different attitudes towards, conceptions of and perspectives on climate change. They include the deep ecology of a German scientist encountered on the voyage, the chants of a Native American woman who was murdered during the massacre at Sand Creek in 1864, the song of a girl from the future and the musings of 'an alien from outer space dressed up in disguise as a fairly amiable dude singin a song written by our – uh – protagonist there, who intended for her song to be sung by a 79 year old climate scientist named Wally' (Hopkins 2013, 20).

On one level the piece is a narrative account of Hopkins' Arctic voyage in the tradition of the Victorian explorers of the nineteenth century. However, the way that this account is woven into a fabric of other different kinds of account also develops a critique of the voyage narrative and pushes beyond the sublime through its disjunctive forms. There is the slightly amateurish and idiosyncratic film documentary of the voyage with short clips of different people she shared the voyage with or happened to meet. This is framed though by the interventions of her cast of fictional characters from the past and the future who offer ghostly commentaries on the eco-political context of the Cape Farewell project. These commentaries are interspersed with Hopkins' autobiographical storytelling and performances of original songs composed in her distinctive orchestrated folk-rock style, reminiscent of singer-songwriters like Joni Mitchell, Ani Di Franco or Natalie Merchant, backed by a small choir performing live, but also integrated within her theatrical frame and sometimes sung in one of her characters. She plays theatrically with her voice and various ways of mediating or remediating it like a low-tech Laurie Anderson. The almost comically chaotic but also coherent mixture of autobiography, science fact and science fiction is characteristic of much of Hopkins' work. She has previously explored cosmology, amnesia and addiction among other complex themes. Her theatrical exploration of her own life experience has integrated music and storytelling with scientific lectures, fictional journeys into space and invented aliens.[1] She is therefore well equipped for navigating between her own sensory experience of a voyage to the Arctic and the accounts of the scientists and other experts she encounters.

Hopkins' piece certainly seems to give an account at a human scale, in so far as this might be understood as autobiography, appearing as herself and telling stories about her own life. Hopkins' work always seems to have an aching honesty, appealing to an idea of shared humanity: she bares her soul. This is twisted though in the last filmed documentary clip she shows in *This Clement World*. Here she bares her body as she walks nude across the snow, in a gesture that seems to echo the exploration of physiological vulnerability by Siobhan Davies I have discussed above. She goes on to reflect on her sense of her own animality and on the planet's surprising clemency towards the human species. The sequence though is also a meditation on the problem of scale and its implications for knowledge:

> In the Arctic it's difficult to gauge distance and scale, because there are almost no reference objects of known size: one might feel one is right up against a glacier when in fact it's miles away. Only in relation to objects of known size, such as birds flying past the glacier, can its massive size be grasped. (Hopkins 2013, 21)

Grasping scale in the Arctic requires juxtapositions with objects of known size. One difficulty with cosmopolitical problems is the apparent remoteness or distance that its objects appear to be from us.[2] Such an object may be the glacier in the Arctic, a sub-atomic particle or a particular molecule in a human cell. The scale of cosmopolitical problems may mean that appropriate 'reference objects' are hard to find. Scientific accounts of these objects may leave them too remote or rely on reference objects that are equally remote. Aesthetic categories like the sublime can hide assumptions about distance and proximity and about its reference objects. In her walking performance she offers herself as 'reference object'. At the same time she reflects on the proliferation of objects near at hand that might obscure a sense of cosmopolitical scale: 'paying our taxes, maximizing our profits, raising our children, coming up with last month's rent' (Hopkins 2013, 21).

Hopkins' approach to cosmopolitical scale positions the work in relation to climate science, to colonial histories and other ways of knowing. She explains at the start of the show why she has integrated screening clips of documentary footage shot by her, with songs and other theatrical sequences:

> [t]his documentary process is my way of building a bridge or opening a doorway between my previously solipsistic existence and a larger world of concern. (Hopkins 2013, 4)

The politics of her practice therefore exemplifies the kind of knowledge repair I have been searching for throughout this book. This is partly to

Figure 6.1 Cynthia Hopkins standing in the doorway in *This Clement World*

do with opening doorways between different temporal and spatial scales. The stage design includes a doorway downstage right (see Figure 6.1) which enables Hopkins' entrances and exits from the stage, her transformations between characters and also appearances from other times and places via a carefully focussed projection. She goes back in history connecting the damage to the planet inflicted by humans, with the devastation to indigenous societies by colonialism. She goes forward into the future to speculate from a perspective after imagined catastrophes to come: the girl sent as 'beneficial emissary' from 200 years into the future has only seen New York City 'through the glass bottom of a boat'. She invents the 'amiable dude from outer space' to gain perspective from another world. She delves into her own 'private melodrama' to explore how far her own depression and addiction to alcohol provide metaphors for broader tendencies towards self-destruction. And yet these insights are not, neither are they intended to be, revelatory. She doesn't presume to have the last word or indeed to offer special insights. She reflects, for instance, on the limitations of the addiction metaphor:

> An alcoholic hits bottom, even if she's ruined everything in her life – which is the only reason she's willing to consider giving up drinking – she can still start over again, no matter how old she is. The human race, on the other hand: ya'all wait until you hit bottom to switch over to renewable energy sources, it's gonna be too late! So, the metaphor don't work. (Hopkins 2013, 20)

Her turn to documentary does not seem to be according to an ethics of authenticity; if anything it is a turn away from her concern with (authentic) autobiographical storytelling that she refers to as solipsistic.

As I have already commented, she intersperses documentary footage with monologues by fantastical characters, her own reflections and songs. The presentation of the documentary sequences also adopts a particular, relatively unusual theatrical convention. She does not give voice to audio recordings as in forms of documentary theatre. Neither does she lip-synch to the original audio. Rather she dubs the film footage in voices that mimic, even gently caricature, the speaker. She actually disavows any particular ethical or aesthetic purpose to her approach in the piece claiming that her approach to dubbing is just the consequence of her poor sound recording skills:

> I've never made a documentary before, and both the image and sound quality of my footage suffered from my lack of experience. The audio recording of the voices in particular was terrible, almost incomprehensible! (Hopkins 2013, 10)

The effect though is to present a series of monologues that speak to and across each other, acknowledge their different sources and different claims or frames as knowledge, but also occupy the same theatrical space.

I have argued elsewhere when writing about the activist artists James Leadbeater and Kota Takeuchi, that this kind of knowing might constitute a kind of amateur science (Parry 2017). Such an idea conceives of the practice as passionate, emerging from love and care, resistant to, or at least reflexive towards, the performance pressures of professionalism. Hopkins is pursuing a related path. Knowing here is not to do with accurate or authentic representation of a climate scientist or an imagined alien from the future. Hopkins is rather evolving approaches at commoning her sense of the world and a relation to it, which she conceives in terms of clemency. To do this she relies on complex processes of assembling collectives: Cape Farewell's eclectic ship's crew, her cast of fantastic characters, film documentary, musicians, songs, audiences. These processes provide mitigated access to a still limited range of ways of knowing: autobiography, histories, climate science, walking, documentary, song, theatre. In their juxtaposition though, such ways of knowing provide reference objects for each other, a bit like Theatre of Debate's process of overflowing frames. They provide constant reminders of the limitations of any one way of knowing through the sensory conflict felt as friction between them. Perhaps more than amateur science, her theatrical approach is characterised by her creative yet humble diplomacy on behalf of those for whom she speaks. She is not setting herself up as an alternative expert, even an amateur one, but rather as a diplomat.

Isabelle Stengers has called for diplomacy as a response to the challenges of cosmopolitics. The risks of diplomacy are always in the potential to betray either your own people or other diplomats, both of whom the diplomat has duties towards. Betrayal on some level is inevitable, but the diplomat can also avoid becoming cynical or indifferent. Diplomats can often be accused of circumlocution or evasiveness:

> [t]he figure of the diplomat, always suspect, accused of evading 'the' issue ('it depends…') or betrayal, always trying to escape static disjunction ('you're either with us or against us'), is interesting precisely because it avoids any romantic attraction for an 'elsewhere' that might inspire the hope of escaping from 'here' once and for all. (Stengers 2011b, 410)

The challenge is both to produce knowledge at the same time as acknowledging its reproduction, co-production and contingency on a particular mode of production. That is to say, to acknowledge the basis on which sense is commoned.

One of Hopkins' characters is the ghost of a Native American woman. In character as this woman, Hopkins sings *In A Sacred Manner I Live*, a song recorded by Frances Densmore (1992) in the early twentieth century in her work documenting Sioux music and culture. The song and the lyric itself have a legacy as an alternative, environmental source of wisdom in a number of books that draw on the culture of the indigenous peoples of the Great Plains (Philip 2005; St. Pierre 1995). However, this appeal to indigenous wisdom is also framed by an account of the ghostly woman's rape and murder at the infamous Sand Creek Massacre. She is in an 'afterworld', but we are also living in the 'afterworld' of this massacre. Hopkins adds further layers to this framing through the way she overwrites it and indeed overlays her body onto other bodies. Throughout the piece Hopkins switches between use of microphone, recording and live voice. She dubs the videos; she speaks and sings; she moves in and out of characters. She appears projected on screens and in person, and plays with relationships between the two. It is a bravura performance of translation. When she appears in costume as the Native American woman, she emerges from the doorway on which she has been projected as herself singing a song, which she overdubs live in performance. She moves in front of a screen on which her hand is projected writing text that appears to be a letter from the ghost of the woman to us, the audience. The text acknowledges that 'someone else's hand is writing this'. The woman no longer has a body and no longer has a voice 'except to sing with every now and then' (Hopkins 2013, 2). The apocalypse for those massacred at Sand Creek has already happened. Some people's clement world is other people's afterworld. 'Good riddance' to 'your civilization' she says as she imagines empires crumbling. We are living in its ruins: the afterworld. Hopkins finds in her adoption of the character a reference point that acknowledges its own appropriations. An imagined afterworld provides a reference object for an imagined apocalypse to come. A mediated letter from the past is a reference object for the reports of climate change. Hopkins inhabits the body of a diplomat from the past to the present as a point of reference for her diplomacy on behalf of the climate scientists she goes on to inhabit. The imagined knowledge of the murdered Native American woman need not be held in simple opposition to the knowledge produced by climate science or even that produced within the mainstream currents of contemporary Western 'civilization'. Rather, different ways of knowing need to offer each other reference points. Creating encounters to make this possible is the work of the diplomat.

I hope in the course of writing this book that I have sustained a reasonably diplomatic approach. Stengers warns against the risk of the careless chemist who, in searching to understand a particular material, ends up dissolving what interests him (Stengers 2011b, 410). My intention has been neither to dissolve science into performance, nor vice versa, nor indeed to dissolve both into politics. In her work Stengers tries to identify a 'psychosocial type' from a series of scientists whom she argues might offer a glimpse of how a 'cosmopolitical Parliament' might be composed. In this book I have been pursuing her speculative approach in searching for a certain psychosocial quality in theatrical processes, in the hope that they might help navigate the rough seas of technoscience and contribute to creating time and space for cosmopolitics.

Notes

1 There are a number of scholarly reviews of Hopkins's *Accidental Trilogy*, a series of performance works created 2004–9 (Alker 2010; Bauerlein 2005; Luber 2010).

2 Timothy Morton has influentially tackled the philosophical, political and aesthetic challenges of scale and its relationship to climate change in his book *Hyperobjects* (2013).

REFERENCES

Abbate, Carolyn. 2001. *In Search of Opera*. Princeton: Princeton University Press.

Adams, Zoë. 2018. '"I Don't Know Why Man's Calling Me Family All of a Sudden": Address and Reference Terms in Grime Music'. *Language and Communication* 60: 11–27. https://doi.org/10.1016/j.langcom.2018.01.004.

Advisory Council on the Misuse of Drugs. 2005. 'Khat Report 2005'. www.gov.uk/government/publications/khat-report-2005--6. Accessed 26 November 2019.

———. 2013. 'Khat Report 2013'. www.gov.uk/government/publications/khat-report-2013. Accessed 26 November 2019.

Afrax, Maxamed Daahir. 2013. 'Between Continuity and Innovation: Transitional Nature of Post-Independence Somali Poetry and Drama, 1960s – the Present'. PhD, London: SOAS, University of London. https://eprints.soas.ac.uk/18432/.

Agamben, Giorgio. 1998. *Homo Sacer: Sovereign Power and Bare Life*. Translated by Daniel Heller-Roazen. Stanford, CA: Stanford University Press.

Alberge, Dalya. 2012. 'Theatre Turns to Facebook to Bring a Younger Audience through the Doors'. *Guardian*, 21 April, sec. Stage. www.theguardian.com/stage/2012/apr/22/theatre-turns-to-facebook.

Alker, Gwendolyn. 2010. 'Stretching the Truth Way into the Future: Cynthia Hopkins's The Success of Failure (Or, The Failure of Success)'. *TDR/The Drama Review* 54 (1): 167–71. https://doi.org/10.1162/dram.2010.54.1.167.

Ancona, Matthew d'. 2017. *Post-Truth: The New War on Truth and How to Fight Back*. London: Ebury Press.

Andrzejewski, B. W. 2011. 'Modern and Traditional Aspects of Somali Drama'. *Journal of African Cultural Studies* 23 (1): 85–95. https://doi.org/10.1080/13696815.2011.581461.

Angliss, Sarah. 2018. 'When Algorithms Meet Machines'. In *The Oxford Handbook of Algorithmic Music*, edited by Roger T. Dean and Alex McLean, 321–4. www.oxfordhandbooks.com/view/10.1093/oxfordhb/9780190226992.001.0001/oxfordhb-9780190226992-e-30?rskey=D1r10X&result=1.

Angliss, Sarah. n.d. 'The Effect: National Theatre'. www.sarahangliss.com/theeffectnt. Accessed 22 June 2018.

Archer, Louise, Jennifer Dewitt, Jonathan Osborne, Justin Dillon, Beatrice Willis, and Billy Wong. 2012. 'Science Aspirations, Capital, and Family Habitus: How Families Shape Children's Engagement and Identification with Science'. *American Educational Research Journal* 49 (5): 881–908. https://doi.org/10.3102/0002831211433290.

Arends, Bergit, and Verity Slater, eds. 2004. *Talking Back to Science: Art, Science and the Personal*. London: The Wellcome Trust.

Arends, Bergit, and Davina Thackara, eds. 2003. *Experiment: Conversations in Art and Science*. London: The Wellcome Trust.

Armstrong, Franny. 2009. *The Age Of Stupid*. Journeyman Pictures.

Atwood, Margaret. 2012. *In Other Worlds: SF and the Human Imagination*. London: Virago.

Auslander, Philip. 2002. 'Live from Cyberspace: Or, I Was Sitting at My Computer This Guy Appeared He Thought I Was a Bot'. *PAJ: A Journal of Performance and Art* 24 (1): 16–21. muse.jhu.edu/article/25829.

———. 2008. *Liveness: Performance in a Mediatized Culture*. London; New York: Routledge.

Baccolini, Raffaella. 2004. 'The Persistence of Hope in Dystopian Science Fiction'. *PMLA: Modern Language Association* 119 (3): 518–21. https://doi.org/10.1632/003081204X20587.

———. 2011. 'Realizing Better Futures, Strong Thought for Hard Times'. In *Utopia Method Vision: The Use Value of Social Dreaming*, edited by Tom Moylan and Raffaella Baccolini. Ralahine Utopian Studies v. 1. Limerick: Peter Lang.

Bachelard, Gaston. 1984. *The New Scientific Spirit*. Boston: Beacon Press.

Baldwin, Nicola. 1995. 'The Gift'. Unpublished script.

Ball, Steve. 1993. 'Theatre in Health Education'. In *Learning through Theatre: New Perspectives on Theatre in Education*, edited by Tony Jackson, 227–38. London: Routledge.

Barrett, Michèle, and Bobby Baker, eds. 2007. *Bobby Baker: Redeeming Features of Daily Life*. London; New York: Routledge.

Barry, Andrew, Georgina Born, and Gisa Weszkalnys. 2008. 'Logics of Interdisciplinarity'. *Economy and Society* 37 (1): 20–49. https://doi.org/10.1080/03085140701760841.

Barsh, Gregory S. 2003. 'What Controls Variation in Human Skin Color?' *PLoS Biology* 1 (1). https://doi.org/10.1371/journal.pbio.0000027.

Bartleet, Carina. 2013. 'Performing Science (Not Ethics) in Analogue's *2401 Objects*'. *Interdisciplinary Science Reviews* 38 (4): 351–64. https://doi.org/10.1179/0308018813z.00000000064.

Bartlett, Mike. 2010. *Earthquakes in London*. London: Methuen Drama.

Bataille, Georges, and Annette Michelson. 1986. 'Un-Knowing: Laughter and Tears'. *October* 36: 89–102. https://doi.org/10.2307/778556.

Bates, Gillian P. 2005. 'History of Genetic Disease: The Molecular Genetics of Huntington Disease – a History'. *Nature Reviews. Genetics* 6 (10): 766–73. https://doi.org/10.1038/nrg1686.

Bauerlein, Judy. 2005. 'Cynthia Hopkins, "Accidental Nostalgia"'. *Theatre Forum* 27: 3–7.

Baum, L. Frank. 2019. *The Wizard of Oz*. London: Puffin Classics.

Beggs, Anne. 2010. '"For Urinetown Is Your Town …": The Fringes of Broadway'. *Theatre Journal* 62 (1): 41–56. www.jstor.org/stable/40587441.

Bennett, Jane. 2001. *The Enchantment of Modern Life: Attachments, Crossings, and Ethics*. Princeton: Princeton University Press.

———. 2010. *Vibrant Matter: A Political Ecology of Things*. Durham, NC: Duke University Press.

Bennett, Susan. 1997. *Theatre Audiences*. London; New York: Routledge.

Bergson, Henri. 2010. *Laughter: An Essay on the Meaning of the Comic*. Translated by Cloudesley Brereton and Fred Rothwell. Digireads.

Berlant, Lauren, and Sianne Ngai. 2016. 'Comedy Has Issues'. *Critical Inquiry* 43 (2): 233–49. https://doi.org/10.1086/689666.

Blair, Rhonda, and Amy Cook, eds. 2016. *Theatre, Performance and Cognition: Languages, Bodies and Ecologies*. London: Methuen.

Bolaki, Stella. 2011. 'Re-Covering the Scarred Body: Textual and Photographic Narratives of Breast Cancer'. *Mosaic: A Journal for the Interdisciplinary Study of Literature* 44 (2): 1–17. www.muse.jhu.edu/article/440539.

Boler, Megan, and Elizabeth Davis. 2018. 'The Affective Politics of the "Post-Truth" Era: Feeling Rules and Networked Subjectivity'. *Emotion, Space and Society* 27 (May): 75–85. https://doi.org/10.1016/j.emospa.2018.03.002.

Born, Georgina, and Andrew Barry. 2010. 'Art-Science'. *Journal of Cultural Economy* 3 (1): 103–19. https://doi.org/10.1080/17530351003617610.

Borry, Pascal, Paul Schotsmans, and Kris Dierickx. 2005. 'The Birth of the Empirical Turn in Bioethics'. *Bioethics* 19 (1): 49–71. https://doi.org/10.1111/j.1467-8519.2005.00424.x.

Bottoms, Stephen, ed. 2007. *Small Acts of Repair: Performance, Ecology and Goat Island*. London; New York: Routledge.

———. 2012. 'Climate Change "Science" on the London Stage'. *Wiley Interdisciplinary Reviews-Climate Change* 3 (4): 339–48. https://doi.org/10.1002/wcc.173.

Bourdieu, Pierre. 1977. *Outline of a Theory of Practice*. Cambridge: Cambridge University Press.

———. 1984. *Distinction: A Social Critique of the Judgement of Taste*. London: Routledge.

Bown, Abi. 2012. 'Dayglo'. Unpublished script.

Boxall, Simon. 2006. 'Journal 14/09/2004'. In *Burning Ice: Art and Climate Change*, edited by David Buckland, Ali MacGilp, and Sion Parkinson, 27. London: Cape Farewell.

Boyce, Niall. 2012. 'Uh-Oh, Love Comes to Town (Product Review)'. *The Lancet* 380 (9857): 1902. https://doi.org/10.1016/S0140-6736(12)62101-1.

Boyd, Michelle. 2010. 'Alto on a Broomstick: Voicing the Witch in the Musical Wicked'. *American Music* 28 (1): 97–118. https://doi.org/10.5406/americanmusic.28.1.0097.

Brah, Avtar. 1996. *Cartographies of Diaspora: Contesting Identities*. London; New York: Routledge.

Bramwell, Richard. 2015. *UK Hip-Hop, Grime and the City: The Aesthetics and Ethics of London's Rap Scenes*. London: Routledge.

Brecht, Bertolt. 2001. *Life of Galileo*. Translated by John Willett. Edited by John Willett and Ralph Manheim. London: Methuen Drama.

Buckland, David, Ali MacGilp, and Sion Parkinson, eds. 2006. *Burning Ice: Art and Climate Change*. London: Cape Farewell.

Callon, Michel. 1986. 'Some Elements of a Sociology of Translation: Domestication of the Scallops and the Fishermen of St Brieuc Bay'. In *Power, Action and Belief: A New Sociology of Knowledge?*, edited by John Law, 196–223. London: Routledge & Kegan Paul.

———. 1998. 'An Essay on Framing and Overflowing: Economic Externalities Revisited by Sociology'. *The Sociological Review* 46: 244–69. https://doi.org/10.1111/j.1467-954X.1998.tb03477.x.

Campuzano, V., L. Montermini, M. D. Moltò, *et al.* 1996. 'Friedreich's Ataxia: Autosomal Recessive Disease Caused by an Intronic GAA Triplet Repeat Expansion'. *Science* 271 (5254): 1423–7. https://doi.org/10.1126/science.271.5254.1423.

Canguilhem, Georges. 1994. *A Vital Rationalist: Selected Writings from Georges Canguilhem*. New York: Zone Books.

Čapek, Karel. 2015. *R.U.R.* Translated by Peter Majer and Cathy Porter. London: Bloomsbury.

Case, Sue-Ellen. 2007. *Performing Science and the Virtual*. London: Routledge.

Castañeda, Claudia, and Lucy Suchman. 2013. 'Robot Visions'. *Social Studies of Science* 44 (3): 315–41. https://doi.org/10.1177/0306312713511868.

Cavalcanti, Ildney. 2000. 'Utopias of/f Language in Contemporary Feminist Literary Dystopias'. *Utopian Studies* 11 (2): 152–80. https://englishtuitionlondon.files.wordpress.com/2010/08/ildney-cavacanti-on-language-in-feminist-literary-dystopias.pdf.

Cavendish, Dominic. 2016. 'Wicked: Why the Musical Defied the Odds and Is Still a Phenomenon 10 Years On'. *Telegraph*, 26 September. www.telegraph.co.uk/theatre/what-to-see/wicked-why-the-musical-defied-the-odds-and-is-still-a-phenomenon/.

Central YMCA. 2016. 'Changing Lives for Nearly Two Centuries'. Central YMCA. 2 June 2016. www.ymca.co.uk/about/feature/happy-healthy-and-more-fulfilled-our-journey-through-ages. Accessed 28 October 2019.

Chadwick, Ruth, and Duncan Wilson. 2018. 'The Emergence and Development of Bioethics in the UK'. *Medical Law Review* 26 (2): 183–201. https://doi.org/10.1093/medlaw/fwy011.

Chakrabarty, Dipesh. 2012. 'Postcolonial Studies and the Challenge of Climate Change'. *New Literary History* 43 (1): 1–18. https://doi.org/10.1353/nlh.2012.0007.
Chapman, Louisa. 1999. 'Speculation, Stringent Reasoning, and Science'. *Bulletin of the World Health Organization* 77 (1): 68–70. www.ncbi.nlm.nih.gov/pmc/articles/PMC2557575/pdf/10206762.pdf (fifth article).
Chaudhuri, Una. 1994. '"There Must Be a Lot of Fish in That Lake": Toward an Ecological Theater'. *Theater* 25 (1): 23–31. https://doi.org/10.1215/01610775-25-1-23.
Churchill, Caryl. 2003. *Far Away*. London: Nick Hern Books.
Clinton, William J. 1997. 'State of the Union Address'. State of the Union. 4 February 1997. http://stateoftheunion.onetwothree.net/texts/19970204.html.
Collins, Harry, and Trevor Pinch. 1998. *The Golem at Large: What You Should Know about Technology*. Cambridge: Cambridge University Press.
Crease, Robert P. 1993. *The Play of Nature: Experimentation as Performance*. Bloomington: Indiana University Press.
Crossley, Mark, and James Yarker. 2017. *Devising Theatre with Stan's Cafe*. London; New York: Bloomsbury Methuen Drama.
C-SPAN. 2015. 'Donald Trump Campaign Rally in Hilton Head, South Carolina'. C-SPAN. 30 December 2015. www.c-span.org/video/?402610–1/donald-trump-campaign-rally-hilton-head-south-carolina. Accessed 28 June 2019.
Daubigny, Pierre, and Bruno Latour. 2018. 'Gaia Global Circus'. Translated by Jim O'Hagan and Griffin McInnes. *Theater* 48 (2): 103–23. https://doi.org/10.1215/01610775-4163026.
Davies, Siobhan. 2006. 'Siobhan Davies'. In *Burning Ice: Art and Climate Change*, edited by David Buckland, Ali MacGilp, and Sion Parkinson, 86–9. London: Cape Farewell.
Davis, Evan. 2017. *Post-Truth: Why We Have Reached Peak Bullshit and What We Can Do About It*. London: Little, Brown.
Dennis, Michael A. 2013. 'Vannevar Bush and US Science Policy'. In *States of Knowledge: The Co-Production of Science and the Social Order*, edited by Sheila Jasanoff, 225–53. London; New York: Routledge.
Densmore, Frances. 1992. *Teton Sioux Music and Culture*. Lincoln: University of Nebraska.
Desai, Sunita, and Anupam B. Jena. 2016. 'Do Celebrity Endorsements Matter? Observational Study of BRCA Gene Testing and Mastectomy Rates after Angelina Jolie's New York Times Editorial'. *British Medical Journal* 355 (December): i6357. https://doi.org/10.1136/bmj.i6357.
Dijck, José van. 1998. *Imagenation: Popular Images of Genetics*. New York: Springer.
Dixon, Steve. 2007. *Digital Performance: A History of New Media in Theater, Dance, Performance Art, and Installation*. Cambridge, MA; London: MIT.
Djerassi, Carl. 2002. 'Contemporary "Science-in-Theatre": A Rare Genre'. *Interdisciplinary Science Reviews* 27 (3): 193–201. https://doi.org/10.1179/030801802225003268.
Dobie, Thomas G. 2019. 'Motion Sickness'. In *Motion Sickness: A Motion Adaptation Syndrome*, edited by Thomas G. Dobie, 1–32. Cham: Springer International Publishing.
Dolan, Jill. 2005. *Utopia in Performance: Finding Hope at the Theater*. Ann Arbor: University of Michigan Press.
Dostoevsky, Fyodor. 1869. *The Idiot*. Translated in 2004 by David McDuff. London: Penguin.
Drucker, Peter F. 1971. *The Age of Discontinuity: Guidelines to Our Changing Society*. London: Pan Books.
Dvoskin, Michelle. 2011. 'Audiences and Critics'. In *The Oxford Handbook of The American Musical*, edited by Raymond Knapp, Mitchell Morris, and Stacy Wolf, 365–77. Oxford: Oxford University Press.
Dyer, Richard. 1992. *Only Entertainment*. London: Routledge.
Ede, Siân. 2004. *Art and Science*. London: I.B. Tauris.
Edson, Margaret. 2000. *Wit*. London: Nick Hern Books.
Educational Theatre Association. 1992. 'The Play Survey'. *Dramatics*, October 1992.
———. 1994. 'The Play Survey'. *Dramatics*, November 1994.
Edwards, David. 2009. *Artscience*. Cambridge, MA: Harvard University Press.
Emmott, Stephen. 2013. *Ten Billion*. London: Penguin.

English, Daylanne K., and Alvin Kim. 2013. 'Now We Want Our Funk Cut: Janelle Monáe's Neo-Afrofuturism'. *American Studies* 52 (4): 217–30. https://doi.org/10.1353/ams.2013.0116.
Epstein, Steven. 1996. *Impure Science: AIDS, Activism, and the Politics of Knowledge*. Berkeley: University of California Press.
Fahy, Declan. 2015. *New Celebrity Scientists: Out of the Lab and Into the Limelight*. Lanham: Rowman & Littlefield.
Farber, Sidney, Louis K. Diamond, Robert D. Mercer, Robert F. Sylvester, and James A. Wolff. 1948. 'Temporary Remissions in Acute Leukemia in Children Produced by Folic Acid Antagonist, 4-Aminopteroyl-Glutamic Acid (Aminopterin)'. *New England Journal of Medicine* 238 (23): 787–93. https://doi.org/10.1056/NEJM194806032382301.
Federici, Silvia. 2014. *Caliban and the Witch*. New York: Autonomedia.
Féral, Josette, and Ronald P. Bermingham. 2002. 'Theatricality: The Specificity of Theatrical Language'. *SubStance* 31 (2): 94–108. https://doi.org/10.1353/sub.2002.0026.
Finegold, Peter, Jo Hutchinson, and Simon Parry. 2014. 'Theatre of Debate Evaluation 2011–2014'. Unpublished report.
Fisher, Mark. 2009. *Capitalist Realism: Is There No Alternative?* Winchester: O Books.
———. 2013a. 'The Secret Sadness of the 21st Century: Mark Fisher Recommends James Blake's Overgrown'. *Telekom Electronic Beats* (blog). 18 April 2013. www.electronicbeats.net/mark-fisher-recommends-james-blakes-overgrown/. Accessed 28 June 2019.
———. 2013b. 'The Metaphysics of Crackle: Afrofuturism and Hauntology'. *Dancecult: Journal of Electronic Dance Music Culture* 5 (2): 42–55. https://doi.org/10.12801/1947-5403.2013.05.02.03.
———. 2016. *The Weird and the Eerie*. London: Repeater.
Foucault, Michel. 1979. *The History of Sexuality*. London: Allen Lane.
———. 2001. *The Order of Things: Archaeology of the Human Sciences*. London: Routledge.
———. 2003. *The Birth of the Clinic: An Archaeology of Medical Perception*. Translated by A. M. Sheridan. London: Routledge Classics.
Fox Umbrellas Ltd. 2014. 'History of Fox Umbrellas'. www.foxumbrellas.com/content/9-fox-umbrellas-ltd. Accessed 28 June 2019.
Franklin, Sarah, Celia Lury, and Jackie Stacey. 2000. *Global Nature, Global Culture*. London: Sage. http://ebookcentral.proquest.com/lib/manchester/detail.action?docID=1204085. Accessed 28 October 2019.
———. 2000. 'Life Itself: Global Nature and the Genetic Imaginary'. In *Global Nature, Global Culture*, edited by Sarah Franklin, Celia Lury, and Jackie Stacey, 188–227. London: SAGE Publications. http://ebookcentral.proquest.com/lib/manchester/detail.action?docID=1204085. Accessed 28 October 2019.
Fraser, Nancy. 1990. 'Rethinking the Public Sphere: A Contribution to the Critique of Actually Existing Democracy'. *Social Text*, no. 25/26: 56–80. https://doi.org/10.2307/466240.
Frayn, Michael. 1998. *Copenhagen*. London: Methuen Drama.
Frenette, Paul S., and George F. Atweh. 2007. 'Sickle Cell Disease: Old Discoveries, New Concepts, and Future Promise'. *Journal of Clinical Investigation* 117 (4): 850–8. https://doi.org/10.1172/JCI30920.
Freshwater, Helen. 2009. *Theatre & Audience*. Basingstoke: Palgrave Macmillan.
Giannachi, Gabriella, and Nigel Stewart. 2005. *Performing Nature: Explorations in Ecology and the Arts*. Oxford: Lang.
Gilderson, Claire. 2019. 'Celebrating 175 Years'. *YMCA Club* (blog). 4 June 2019. www.ymcaclub.co.uk/blog/celebrating-175-years. Accessed 28 October 2019.
Giroux, Henry A. 2001. 'Cultural Studies as Performative Politics'. *Cultural Studies: Critical Methodologies* 1 (1): 5–23. https://doi.org/10.1177/153270860100100102.
Goffman, Erving. 1986. *Frame Analysis: An Essay on the Organization of Experience*. Boston: Northeastern University Press.
Goodall, Jane. 2002. *Performance and Evolution in the Age of Darwin: Out of the Natural Order*. London; New York: Routledge.

Gould, Deborah B. 2009. *Moving Politics: Emotion and ACT UP's Fight against AIDS*. Chicago: University of Chicago Press.
Gregory, Fiona. 2016. 'Hybrid Creatures: Mrs Patrick Campbell's Contributions to Pygmalion'. *Nineteenth Century Theatre and Film* 43 (1): 107–21. https://doi.org/10.1177/1748372716650204.
Gregory, Jane, and Steve Miller. 2000. *Science in Public: Communication, Culture, And Credibility*. Cambridge, MA: Perseus Publishing.
Griffiths, Trevor. 1977. *Through the Night and Such Impossibilities*. London: Faber & Faber.
Guggenheim, Davis. 2006. *An Inconvenient Truth*. Paramount Home Entertainment.
Gumperz, John Joseph. 1982. *Discourse Strategies*. Cambridge: Cambridge University Press.
Habermas, Jürgen. 1992. *The Structural Transformation of the Public Sphere: Inquiry into a Category of Bourgeois Society*. Cambridge: Polity Press.
Haley, Jennifer. 2014. *The Nether*. London: Faber & Faber.
Hall, J. M., M. K. Lee, B. Newman, J. E. Morrow, L. A. Anderson, B. Huey, and M. C. King. 1990. 'Linkage of Early-Onset Familial Breast Cancer to Chromosome 17q21'. *Science* 250 (4988): 1684–9. https://doi.org/10.1126/science.2270482.
Haraway, Donna. 1991. *Simians, Cyborgs and Women: The Reinvention of Nature*. London: Free Association Books.
———. 1997. *Modest_Witness@Second_Millennium.FemaleMan_Meets_OncoMouse: Feminism and Technoscience*. New York; London: Routledge.
Harjunpää, Katariina, and Aleksi Mäkilähde. 2015. 'Reiteration: At the Intersection of Code-Switching and Translation'. *Multilingua* 35 (2): 163–201. https://doi.org/10.1515/multi-2015-0036.
Harman, Graham. 2014. *Bruno Latour: Reassembling the Political*. London: Pluto Press.
———. 2016. *Immaterialism: Objects and Social Theory*. Cambridge: Polity Press.
Harrod, Mark. 2011. 'Values⊠Based Global Governance: The YMCA Story (So Far)'. *Global Policy* 2 (2): 226–8. https://doi.org/10.1111/j.1758-5899.2011.00078.x.
Hawking, Stephen W., and Thomas Hertog. 2018. 'A Smooth Exit from Eternal Inflation?' *Journal of High Energy Physics* 2018 (4): 147. https://doi.org/10.1007/JHEP04(2018)147.
Hayles, N. Katherine. 1999. *How We Became Posthuman: Virtual Bodies in Cybernetics, Literature, and Informatics*. Chicago: University of Chicago Press.
Haynes, Roslynn. 2003. 'From Alchemy to Artificial Intelligence: Stereotypes of the Scientist in Western Literature'. *Public Understanding of Science* 12 (3): 243–53. https://doi.org/10.1177/0963662503123003.
Headlong. n.d. 'About Us'. Headlong Theatre. https://headlong.co.uk/about/. Accessed 5 July 2018.
Herndl, Diane Price. 2014. 'Virtual Cancer: BRCA and Posthuman Narratives of Deleterious Mutation'. *Tulsa Studies in Women's Literature* 33 (1): 25–45. www.jstor.org/stable/43653275.
Heuvel, Edward van den. 2016. 'The Big Bang as Origin of the Universe'. In *The Amazing Unity of the Universe: And Its Origin in the Big Bang*, edited by Edward van den Heuvel, 137–60. Cham: Springer International Publishing. https://doi.org/10.1007/978-3-319-23543-1_9.
Hilgartner, Stephen. 2000. *Science on Stage: Expert Advice as Public Drama*. Stanford: Stanford University Press.
Hirst, Olivia. 2014. 'Goodstock'. Unpublished script.
———. 2015. 'Why a Double Mastectomy before the Age of 30 Can Be a Laughing Matter'. *Guardian*, 12 August, sec. Opinion. www.theguardian.com/commentisfree/2015/aug/12/double-mastectomy-30-cancer-gene-goodstock.
Hochschild, Arlie Russell. 1979. 'Emotion Work, Feeling Rules, and Social Structure'. *American Journal of Sociology* 85 (3): 551–75. https://doi.org/10.1086/227049.
Hodges, Andrew. 2014. *Alan Turing: The Enigma*. New York: Vintage.
Holden, Constance, and Yudhijit Bhattacharjee. 2003. 'Random Samples'. *Science* 301 (5640): 1665–7. www.jstor.org/stable/3835210.
Holzman, Winnie, and Stephen Schwartz. 2006. 'The Songs and Story of Wicked'. In *Wicked: The Grimmerie*, edited by Holly Rothman, 139–77. New York: Hyperion.

Hopkins, Cynthia. 2013. 'This Clement World'. Unpublished script.
House of Lords Science and Technology Select Committee. 2000. 'Science and Society'. 3rd Report. London.
Hulme, Mike. 2009. *Why We Disagree about Climate Change: Understanding Controversy, Inaction and Opportunity*. Cambridge: Cambridge University Press.
Human Fertilisation and Embryology Authority. 2019. 'PGD Conditions'. www.hfea.gov.uk/pgd-conditions/. Accessed 28 October 2019.
Huppert, Doreen, and Thomas Brandt. 2018. 'Dizziness and Vertigo Syndromes Viewed with a Historical Eye'. *Journal of Neurology*, March, 1–7. https://doi.org/10.1007/s00415-018-8807-x.
Huxley, Aldous. 2007. *Brave New World*. London: Vintage Classics.
Ibsen, Henrik. 2008. *An Enemy of the People*. Translated by Rebecca Lenkiewicz and Charlotte Barslund. London: Faber and Faber.
Jackson, Anthony, ed. 1986. *Learning Through Theatre: Essays and Casebooks on Theatre in Education*. 1st edn. Manchester: Manchester University Press.
———. 1993. *Learning through Theatre: New Perspectives on Theatre in Education*. 2nd edn. London: Routledge.
Jackson, Anthony. 2007. *Theatre, Education and the Making of Meanings: Art or Instrument?* Manchester: Manchester University Press.
Jackson, Anthony, and Chris Vine, eds. 2013. *Learning Through Theatre*. 3rd edn. London: Routledge.
Jameson, Fredric. 2007. *Archaeologies of the Future: The Desire Called Utopia and Other Science Fictions*. London: Verso.
Jasanoff, Sheila, ed. 2013. *States of Knowledge: The Co-Production of Science and the Social Order*. London; New York: Routledge.
Jays, David. 2017. 'Kafka, Catwalks and Vanishing Sets: Miriam Buether's Stunning Designs – in Pictures'. *Guardian*, 4 May, sec. Stage. www.theguardian.com/stage/2017/may/04/miriam-buether-stunning-stage-set-designs-in-pictures.
Jolie, Angelina. 2013. 'My Medical Choice'. *The New York Times*, 14 May, sec. Opinion. www.nytimes.com/2013/05/14/opinion/my-medical-choice.html.
Kac, Eduardo, Roger F. Malina, and Sean Cubitt. 2009. *Signs of Life: Bio Art and Beyond*. Cambridge, MA: MIT Press.
Kane, Sarah. 2001. *Complete Plays: Blasted, Phaedra's Love, Cleansed, Crave, 4.48 Psychosis, Skin*. London: Methuen Drama.
Kazzazi, Seyedeh Anahit. 2017. 'The Anatomy of the Science Play'. *New Theatre Quarterly* 33 (4): 333–44. https://doi.org/10.1017/S0266464X17000471.
Kemp, Martin. 2006. *Seen | Unseen: Art, Science, and Intuition from Leonardo to the Hubble Telescope*. Oxford: Oxford University Press.
Kennedy, Joseph V. 2012. 'The Sources and Uses of U.S. Science Funding'. *The New Atlantis*. www.thenewatlantis.com/publications/the-sources-and-uses-of-us-science-funding. Accessed 28 June 2019.
Kershaw, Baz. 2007. *Theatre Ecology: Environments and Performance Events*. Cambridge: Cambridge University Press.
Kimmings, Bryony, Brian Lobel, and Tom Parkinson. 2018. *A Pacifist's Guide to the War on Cancer*. London: Oberon Books.
Klawiter, Maren. 2008. *The Biopolitics of Breast Cancer: Changing Cultures of Disease and Activism*. Minneapolis: University of Minnesota Press.
Knorr-Cetina, Karin D. 1999. *Epistemic Cultures: How the Sciences Make Knowledge*. Cambridge, MA: Harvard University Press.
Kolehmainen, Leena, and Janne Skaffari. 2015. 'Multilingual Practices in Contemporary and Historical Contexts: Interfaces between Code-Switching and Translation'. *Multilingua* 35 (2): 123–35. https://doi.org/10.1515/multi-2015-0038.
Kotis, Greg, and Mark Hollmann. 2014. *Urinetown*. London: Nick Hern Books.
Kubrick, Stanley. 1960. *Spartacus*. Bryna Productions.
Kuhn, Thomas S., and Ian Hacking. 2012. *The Structure of Scientific Revolutions*. Chicago: University of Chicago Press.

Laclau, Ernesto. 2018. *On Populist Reason*. London: Verso.
Latour, Bruno. 1987. *Science in Action: How to Follow Scientists and Engineers through Society*. Cambridge, MA: Harvard University Press.
———. 1988. *The Pasteurization of France*. Cambridge, MA : Harvard University Press.
———. 2004a. 'Why Has Critique Run out of Steam? From Matters of Fact to Matters of Concern'. *Critical Inquiry* 30 (2): 225–48. https://doi.org/10.1086/421123.
———. 2004b. *Politics of Nature: How to Bring the Sciences into Democracy*. Cambridge, MA: Harvard University Press.
Launer, John. 2017. 'Henry Wellcome: The Man Who Made Medicine'. *Postgraduate Medical Journal* 93 (1102): 507–8. https://doi.org/10.1136/postgradmedj-2017-135182.
Lavery, Carl. 2015. 'Ionesco's Green Lesson'. In *Rethinking the Theatre of the Absurd : Ecology, the Environment and the Greening of the Modern Stage*, edited by Carl Lavery and Clare Finburgh, 165–90. London: Bloomsbury.
Lavery, Carl, and Clare Finburgh, eds. 2015. *Rethinking the Theatre of the Absurd : Ecology, the Environment and the Greening of the Modern Stage*. London: Bloomsbury.
Leach, Melissa, Ian Scoones, and Brian Wynne. 2005. *Science and Citizens : Globalization and the Challenge of Engagement*. London: Zed.
Lederer, Susan E. 2007. 'Dark Victory: Cancer and Popular Hollywood Film'. *Bulletin of the History of Medicine* 81 (1): 94–115. www.jstor.org/stable/44451740.
LePage, Louise. 2016. '"Thinking Something Makes It So": Performing Robots, the Workings of Mimesis and the Importance of Character'. In *Twenty-First Century Drama*, 279–301. Basingstoke: Palgrave Macmillan.
Letts, Quentin. 2012. 'First Night Review of The Effect: Billie's Brilliant but the Play Is a Bit of a Trial'. *Mail Online*, 14 November. www.dailymail.co.uk/tvshowbiz/reviews/article-2232654/The-Effect-Royal-National-Theatre-First-night-review-Quentin-Letts.html.
Levin, Alex V. 2016. 'Retinal Hemorrhage: Science versus Speculation'. *Journal of American Association for Pediatric Ophthalmology and Strabismus* 20 (2): 93–5. https://doi.org/10.1016/j.jaapos.2016.03.001.
Levinson, Ralph. 2010. 'Science Education and Democratic Participation: An Uneasy Congruence?' *Studies in Science Education* 46 (1): 69–119. https://doi.org/10.1080/03057260903562433.
Levinson, Ralph, Helen Nicholson, and Simon Parry. 2008. *Creative Encounters: New Conversations in Science Education and the Arts*. London: Wellcome Trust.
Lobel, Brian. 2008. '*BALL*: The Script'. *Text and Performance Quarterly* 28 (1–2): 160–77. https://doi.org/10.1080/10462930701754416.
Loomis, Chauncey. 1977. 'The Arctic Sublime'. In *Nature and the Victorian Imagination*, edited by U. C. Knoepflmacher and G. B. Tennyson, 95–112. Berkeley: University of California Press.
Lovell, Bernard. 2001. 'Obituary: Professor Sir Fred Hoyle'. *Guardian*, 23 August. www.theguardian.com/news/2001/aug/23/guardianobituaries.spaceexploration. Accessed 28 June 2019.
Luber, Steve. 2010. 'The Impossibility of Cynthia Hopkins: Or, Can Autobiographical Musical Performance Save the Space Program??' *PAJ: A Journal of Performance and Art* 32 (1): 76–82. https://doi.org/10.1162/pajj.2010.32.1.76.
Macmillan, Duncan. 2015. *People, Places and Things*. London: Oberon Books.
Marres, Noortje. 2012. *Material Participation: Technology, the Environment and Everyday Publics*. Basingstoke: Palgrave Macmillan.
Marx, Karl, and Friedrich Engels. 2015. *The Communist Manifesto*. London: Penguin Classics.
Massey, Doreen. 1999. 'Imagining Globalization: Power-Geometries of Time-Space'. In *Global Futures: Migration, Environment and Globalization*, edited by Avtar Brah, Mary J. Hickman, and Máirtín Mac an Ghaill, 27–44. Explorations in Sociology. Basingstoke: Palgrave Macmillan.
———. 2005. *For Space*. London: Sage.

Mbembe, Achille. 2017. *Critique of Black Reason*. Translated by Laurent Dubois. Durham, NC: Duke University Press.

Mbembe, Achille, and Libby Meintjes. 2003. 'Necropolitics'. *Public Culture* 15 (1): 11–40. https://warwick.ac.uk/fac/arts/english/currentstudents/postgraduate/masters/modules/postcol_theory/mbembe_22necropolitics22.pdf.

McCrone, P., K. Bhui, T. Craig, S. Mohamud, N. Warfa, S. A. Stansfeld, G. Thornicroft, and S. Curtis. 2005. 'Mental Health Needs, Service Use and Costs among Somali Refugees in the UK'. *Acta Psychiatrica Scandinavica* 111 (5): 351–7. https://doi.org/10.1111/j.1600-0447.2004.00494.x.

McKenzie, Jon. 2001. *Perform or Else: From Discipline to Performance*. New York: Routledge.

Menkhaus, Ken. 2008. 'The Role and Impact of the Somali Diaspora in Peace-Building, Governance and Development'. In *Africa's Finances: The Contribution of Remittances*, edited by Raj Bardouille, Muna Ndulo, and Margaret Grieco, 187–202. Cambridge: Cambridge Scholars Publishing.

Merchant, Carolyn. 2006. 'The Scientific Revolution and The Death of Nature'. *Isis* 97 (3): 513–33. https://doi.org/10.1086/508090.

Mermikides, Alex, and Gianna Bouchard, eds. 2016. *Performance and the Medical Body*. London: Bloomsbury 3PL.

Merton, Robert. 1973. *The Sociology of Science: Theoretical and Empirical Investigations*. Chicago: University of Chicago Press.

Met Office. 2001. 'HadCM3: Met Office Climate Prediction Model'. www.metoffice.gov.uk/research/modelling-systems/unified-model/climate-models/hadcm3. Accessed 11 June 2019.

Meyer-Dinkgräfe, Daniel. 2015. 'Liveness: Phelan, Auslander, and After'. *Journal of Dramatic Theory and Criticism* 29 (2): 69–79. https://doi.org/10.1353/dtc.2015.0011.

Millar, Robin, and Jonathan Osborne. 2000. 'Meeting the Challenge of Change'. *Studies in Science Education* 35 (1): 190–7. https://doi.org/10.1080/03057260008560164.

Miller, Arthur I. 2014. *Colliding Worlds: How Cutting-Edge Science Is Redefining Contemporary Art*. New York: W. W. Norton & Company.

Miller, Greg. 2009. 'FMRI Evidence Used in Murder Sentencing'. *Science Magazine*, 23 November. www.sciencemag.org/news/2009/11/fmri-evidence-used-murder-sentencing. Accessed 28 October 2019.

Mohamud, S., K. Bhui, N. Warfa, S. Stansfeld, and T. Craig. 2002. 'Cross-Cultural Translation and Validation of Mental Health Outcome Measures among Somali Refugees in London'. *Acta Psychiatrica Scandinavica* 105 (s411): 13. https://doi-org.manchester.idm.oclc.org/10.1034/j.1600-0447.2002.105s411011.x.

Monáe, Janelle. 2008. *Janelle Monae: Many Moons* [Official Short Film]. www.youtube.com/watch?v=LHgbzNHVg0c. Accessed 28 June 2019.

Morgan, Benjamin. 2016. 'After the Arctic Sublime'. *New Literary History* 47 (1): 1–26. https://doi.org/10.1353/nlh.2016.0000.

Morton, Timothy. 2013. *Hyperobjects Philosophy and Ecology after the End of the World*. Minneapolis: University of Minnesota Press.

Moura, Leonel. 2010. 'RUR: The Birth of the Robot'. www.leonelmoura.com/index.php/robot-art/rur/. Accessed 28 June 2019.

Moylan, Tom. 2000. *Scraps of the Untainted Sky: Science Fiction, Utopia, Dystopia*. Boulder: Westview Press.

Moylan, Tom, and Raffaella Baccolini, eds. 2011. *Utopia Method Vision: The Use Value of Social Dreaming*. Bern: Peter Lang.

Mukherjee, Siddhartha. 2010. *The Emperor of All Maladies: A Biography of Cancer*. New York: Scribner Book Company.

Muñoz, José Esteban. 2009. *Cruising Utopia: The Then and There of Queer Futurity*. New York: New York University Press.

Musgrave, Ben. 2012. 'Stunted Trees and Broken Bridges'. Unpublished script.

Nahum, Andrew. 2017. 'Automata, Androids and Life'. In *Robots: The 500-Year Quest to Make Machines Human*, edited by Ben Russell, 51–69. London: Scala Arts & Heritage Publishers Ltd.

Needham, Joseph. 2000. *Science and Civilisation in China*. Cambridge: Cambridge University Press.

Ngai, Sianne. 2016. 'Theory of the Gimmick'. *Critical Inquiry* 43 (2): 466–505. https://doi.org/10.1086/689672.

Nicholson, Helen. 2009. *Theatre & Education*. Basingstoke: Palgrave Macmillan.

———. 2011. *Theatre, Education and Performance*. New York: Palgrave.

Nordmann, Alfred. 2010. 'A Forensics of Wishing: Technology Assessment in the Age of Technoscience'. *Poiesis & Praxis* 7 (1–2): 5–15. https://doi.org/10.1007/s10202-010-0081-7.

Noys, Benjamin. 2014. *Malign Velocities: Accelerationism and Capitalism*. Winchester: Zero Books.

Nye, Coleman. 2012. 'Cancer Previval and the Theatrical Fact'. *TDR/The Drama Review* 56 (4): 104–20. https://doi.org/10.1162/DRAM_a_00217.

Orwell, George. 2004. *1984 Nineteen Eighty-Four*. London: Penguin Classics.

Page, Louise. 1990. *Plays: 1: Tissue; Salonika; Real Estate; Golden Girls*. London: Methuen Drama.

Paralympic Games. 2012. *Opening Ceremony: London 2012 Paralympic Games*. www.youtube.com/watch?v=Kd4FgGSY5BY. Accessed 28 June 2019.

Parker-Starbuck, Jennifer. 2011. *Cyborg Theatre: Corporeal/Technological Intersections in Multimedia Performance*. Basingstoke: Palgrave Macmillan.

Parry, Simon, ed. 2005. *Pulse: Youth Arts Inspired by Biomedical Science*. London: Wellcome Trust.

———. 2010. 'Imagining Cosmopolitan Space: Spectacle, Rice and Global Citizenship'. *Research in Drama Education: The Journal of Applied Theatre and Performance* 15 (3): 317–37. https://doi.org/10.1080/13569783.2010.495266.

———. 2017. 'Amateur Science in Activist Performance: Towards a Slow Science'. *Contemporary Theatre Review* 27 (1): 61–75. https://doi.org/10.1080/10486801.2016.1262847.

'Patricia Loughrey – Plays'. n.d. https://patricialoughrey.weebly.com/plays.html. Accessed 8 August 2018.

People, Science and Policy. 2006. 'The Ethics of Brain Research: Evaluation of Mind the Gap'. Unpublished report.

Phelan, Peggy. 1993. *Unmarked: The Politics of Performance*. London; New York: Routledge.

———. 2003. 'Performance, Live Culture and Things of the Heart'. *Journal of Visual Culture* 2 (3): 291–302. https://doi.org/10.1177/14704129030020030002.

Philip, Neil, ed. 2005. *In a Sacred Manner I Live: Native American Wisdom*. New York: HMH Books for Young Readers.

Pickersgill, Martyn. 2011. 'Connecting Neuroscience and Law: Anticipatory Discourse and the Role of Sociotechnical Imaginaries'. *New Genetics and Society* 30 (1): 27–40. https://doi.org/10.1080/14636778.2011.552298.

Pignarre, Philippe, and Isabelle Stengers. 2011. *Capitalist Sorcery: Breaking the Spell*. Translated by Andrew Goffey. Basingstoke: Palgrave Macmillan.

Pinch, Trevor J. 1992. 'Opening Black Boxes: Science, Technology and Society'. *Social Studies of Science* 22 (3): 487–510. www.jstor.org/stable/370279?seq=1.

Popper, Karl R. 2002. *The Logic of Scientific Discovery*. London: Routledge.

Prebble, Lucy. 2012. *The Effect*. London: Methuen Drama.

Press Association. 2014. '"Angelina Effect" More than Doubled Breast Cancer Tests in UK, Study Finds'. *Guardian*, 19 September, sec. Society. www.theguardian.com/society/2014/sep/19/angelina-jolie-breast-cancer-study.

Putney, Clifford. 2001. *Muscular Christianity: Manhood and Sports in Protestant America, 1880–1920*. Cambridge, MA: Harvard University Press.

'Pygmalion, His Majesty's Theatre, 1914, Review'. 2014. *Telegraph*, 11 April, sec. Culture. www.telegraph.co.uk/culture/theatre/theatre-reviews/10757120/Pygmalion-His-Majestys-Theatre-1914-review.html.

Rabinow, Paul. 2002. *French DNA: Trouble in Purgatory*. Chicago: University of Chicago Press.

Racine, Eric, Sarah Waldman, Jarett Rosenberg, and Judy Illes. 2010. 'Contemporary Neuroscience in the Media'. *Social Science & Medicine* 71 (4): 725–33. https://doi.org/10.1016/j.socscimed.2010.05.017.

Ranscombe, Peter. 2013. 'HeLa'. *The Lancet Oncology* 14 (6): 459. https://doi.org/10.1016/S1470-2045(13)70200-6.
Rapley, Chris, and Duncan Macmillan. 2015. *2071: The World We'll Leave Our Grandchildren*. London: John Murray.
Ray, Marcie. 2014. 'My Fair Lady: A Voice for Change'. *American Music* 32 (3): 292–316. www.jstor.org/stable/10.5406/americanmusic.32.3.0292?seq=1.
Rayner, Steve. 2012. 'Uncomfortable Knowledge: The Social Construction of Ignorance in Science and Environmental Policy Discourses'. *Economy and Society* 41 (1): 107–25. https://doi.org/10.1080/03085147.2011.637335.
Rebellato, Dan. 2017. 'Of an Apocalyptic Tone Recently Adopted in Theatre: British Drama, Violence and Writing'. *Sillages Critiques*, no. 22 (March). http://journals.openedition.org/sillagescritiques/4798.
Reid, Kerry. 2018. 'Sideshow's Ambitious "HeLa" Traces the Cosmic Wheels Back to a Woman with Cancer in 1951'. *Chicago Tribune*, 27 November. www.chicagotribune.com/entertainment/theater/ct-ent-hela-sideshow-review-1128-story.html.
Reilly, Kara. 2011. *Automata and Mimesis on the Stage of Theatre History*. Basingstoke: Palgrave Macmillan.
Risner, Doug. 2000. 'Making Dance, Making Sense: Epistemology and Choreography'. *Research in Dance Education* 1 (2): 155–72. https://doi.org/10.1080/713694259.
Rittel, Horst W. J., and Melvin M. Webber. 1973. 'Dilemmas in a General Theory of Planning'. *Policy Sciences* 4 (2): 155–69. https://doi.org/10.1007/BF01405730.
Roco, Mihail C., and William Sims Bainbridge. 2002. 'Converging Technologies for Improving Human Performance: Nanotechnology, Biotechnology, Information Technology and Cognitive Science'. World Technology Evaluation Center. www.wtec.org/ConvergingTechnologies//1/NBIC_report.pdf. Accessed 8 July 2020.
Roman, David. 1998. *Acts of Intervention: Performance, Gay Culture, and AIDS*. Bloomington: Indiana University Press.
Root, Michael. 2003. 'The Use of Race in Medicine as a Proxy for Genetic Differences'. *Philosophy of Science* 70 (5): 1173–83. https://doi.org/10.1086/377398.
Rose, Nikolas. 1999. *Powers of Freedom: Reframing Political Thought*. Cambridge: Cambridge University Press.
Rose, Nikolas. 2007. *Politics of Life Itself: Biomedicine, Power and Subjectivity in the Twenty-First Century*. Princeton: Princeton University Press.
Rotman, Brian. 2008. *Becoming beside Ourselves: The Alphabet, Ghosts, and Distributed Human Being*. Durham, NC: Duke University Press.
Royal Society, The. 1985. *The Public Understanding of Science*. London: The Royal Society.
Russell, Ben. 2017. *Robots: The 500-Year Quest to Make Machines Human*. London: Scala Arts & Heritage Publishers Ltd.
Saini, Angela. 2009. 'The Brain Police: Judging Murder with an MRI'. *Wired UK*, 27 May. www.wired.co.uk/article/guilty.
Sample, Ian, and Alex Hern. 2014. 'Scientists Dispute Whether Computer "Eugene Goostman" Passed Turing Test'. *Guardian*, 9 June, sec. Technology. www.theguardian.com/technology/2014/jun/09/scientists-disagree-over-whether-turing-test-has-been-passed.
Santos, Boaventura de Sousa. 2014. *Epistemologies of the South: Justice Against Epistemicide*. Boulder: Routledge.
Schanze, Jens. 2010. *Plug & Pray*. Farbfilm Home Entertainment.
Schechner, Richard, and Victor Turner. 1985. *Between Theater and Anthropology*. Philadelphia: University of Pennsylvania Press.
Scheer, Edward. 2015. 'Robotics as New Media Dramaturgy: The Case of the Sleepy Robot'. *TDR: The Drama Review* 59 (3): 140–9. https://www.mitpressjournals.org/doi/10.1162/DRAM_a_00476.
Schneider, Rebecca. 2015. 'New Materialisms and Performance Studies'. *TDR: The Drama Review* 59 (4): 7–17. www.mitpressjournals.org/doi/abs/10.1162/DRAM_a_00493?journalCode=dram.
Science Media Centre. n.d. 'About Us'. www.sciencemediacentre.org/about-us/. Accessed 30 November 2018.

Sedgman, Kirsty. 2018. *The Reasonable Audience: Theatre Etiquette, Behaviour Policing, and the Live Performance Experience*. New York: Palgrave Pivot.

Sedgman, Kirsty, and John E. McGrath. 2016. *Locating the Audience: How People Found Value in National Theatre Wales*. Chicago: University of Chicago Press.

Sedgwick, Eve Kosofsky. 2003. *Touching Feeling: Affect, Pedagogy, Performativity*. Durham, NC: Duke University Press.

Senior, Adele. 2011. 'Haunted by Henrietta: The Archive, Immortality and the Biological Arts'. *Contemporary Theatre Review* 21 (4): 511–29. https://doi.org/10.1080/10486801.2011.610306.

Sha, Richard C. 2013. 'Romantic Physiology and the Work of Romantic Imagination: Hypothesis and Speculation in Science and Coleridge'. *European Romantic Review* 24 (4): 403–19. https://doi.org/10.1080/10509585.2013.807966.

Shapin, Steven, and Simon Schaffer. 2011. *Leviathan and the Air-Pump: Hobbes, Boyle, and the Experimental Life*. Princeton: Princeton University Press.

Shaughnessy, Nicola. 2012. *Applying Performance: Live Art, Socially Engaged Theatre and Affective Practice*. Basingstoke: Palgrave Macmillan.

Shaw, George Bernard. 1987a. *The Doctor's Dilemma: A Tragedy*. Edited by Dan Laurence. Harmondsworth: Penguin Classics.

———. 1987b. *Pygmalion: A Romance in Five Acts*. London: Penguin.

Shepherd-Barr, Kirsten. 2006. *Science on Stage: From Doctor Faustus to Copenhagen*. Princeton: Princeton University Press.

———. 2015. *Theatre and Evolution from Ibsen to Beckett*. New York: Columbia University Press.

Shiva, Vandana, Maria Mies, and Ariel Salleh. 2014. *Ecofeminism*. London: Zed Books.

Siedel, Torsten. 2015. *Hybride Steuerung parallel gekoppelter Aktoren am Beispiel des humanoiden Roboters Myon*. Berlin: Humboldt-Universität zu Berlin, Mathematisch-Naturwissenschaftliche Fakultät. https://doi.org/10.18452/17368.

Sierz, Aleks. 2002. 'Still In-Yer-Face? Towards a Critique and a Summation'. *New Theatre Quarterly* 18 (1): 17–24. https://doi.org/10.1017/S0266464X0200012X.

Singh, Manmohan. 2012. 'Statement at the XVI Summit of the Non-Aligned Movement'. *The Hindu*, 30 August. www.thehindu.com/news/resources/statement-by-prime-minister-manmohan-singh-at-the-xvi-summit-of-the-nonaligned-movement/article3839792.ece. Accessed 28 June 2019.

Siobhan Davies Dance. n.d. 'Company History'. Siobhan Davies Dance. www.siobhandavies.com/company-history/. Accessed 13 June 2019.

Skloot, Rebecca. 2011. *The Immortal Life of Henrietta Lacks*. London: Pan.

Snow, C. P. 2008. *The Two Cultures*. Cambridge: Cambridge University Press.

Sofer, Andrew. 2003. *The Stage Life of Props*. Ann Arbor: University of Michigan Press.

Sokal, Alan D. 1996. 'Transgressing the Boundaries: Toward a Transformative Hermeneutics of Quantum Gravity'. *Social Text*, no. 46/47: 217–52. https://pdfs.semanticscholar.org/c883/3dbd65a05b7826e46e1c5a3df2b464184c00.pdf.

———. 1998. *Fashionable Nonsense: Postmodern Intellectuals' Abuse of Science*. New York: Picador.

St. Pierre, Mark. 1995. *Walking in the Sacred Manner: Healers, Dreamers, and Pipe Carriers - Medicine Women of the Plains*. New York: Touchstone.

Stacey, Jackie. 1997. *Teratologies: A Cultural Study of Cancer*. London: Routledge.

Steels, Luc, and Manfred Hild, eds. 2012. *Language Grounding in Robots*. New York: Springer.

Stengers, Isabelle. 2000a. 'Another Look: Relearning to Laugh'. Translated by Penelope Deutscher. *Hypatia* 15 (4): 41–54. www.jstor.org/stable/3810672.

———. 2000b. *The Invention of Modern Science*. Minneapolis: University of Minnesota Press.

———. 2005. 'The Cosmopolitical Proposal'. In *Making Things Public: Atmospheres of Democracy*, edited by Bruno Latour and Peter Weibel, 994–1003. Cambridge, MA: MIT Press.

———. 2010. *Cosmopolitics I*. Minneapolis: University of Minnesota Press.

———. 2011a. 'Wondering about Materialism'. In *The Speculative Turn : Continental Materialism and Realism*, edited by Levi Bryant, Nick Srnicek, and Graham Harman, 368–80. Prahran, Vic.: re.press.

———. 2011b. *Cosmopolitics II*. Translated by Robert Bononno. Minneapolis: University of Minnesota Press.

Stott, Peter A., S. F. B. Tett, G. S. Jones, M. R. Allen, J. F. B. Mitchell, and G. J. Jenkins. 2000. 'External Control of 20th Century Temperature by Natural and Anthropogenic Forcings'. *Science* 290 (5499): 2133–7. https://science.sciencemag.org/content/290/5499/2133.full.

Sturm, Richard A. 2006. 'A Golden Age of Human Pigmentation Genetics'. *Trends in Genetics: TIG* 22 (9): 464–8. https://doi.org/10.1016/j.tig.2006.06.010.

Suchman, Lucy. 2007. *Human-Machine Reconfigurations: Plans and Situated Actions*. Cambridge: Cambridge University Press.

———. 2015. 'Situational Awareness: Deadly Bioconvergence at the Boundaries of Bodies and Machines'. *MediaTropes Test Journal* 5 (1): 1–24. https://doaj.org/article/fc00d7d7fa124fa1af135aa555383a94.

Swash, Rosie. 2011. 'Poly Styrene Dies Aged 53'. *Guardian*, 26 April, sec. Music. www.theguardian.com/music/2011/apr/26/poly-styrene-dies-aged-53.

Tallis, Raymond. 2014. *Aping Mankind: Neuromania, Darwinitis and the Misrepresentation of Humanity*. London; New York: Routledge.

Telegraph. 2015. 'George Bernard Shaw: Theatre-Goers Should Not Laugh Out Loud'. www.telegraph.co.uk/books/authors/george-bernard-shaw-interview/.

Tetteh, Dinah A. 2019. 'Visibility of the Ovaries and Breasts: An Analysis of Angelina Jolie's Medical Announcements'. *Feminist Media Studies* 19 (5): 651–66. https://doi.org/10.1080/14680777.2018.1447979.

The Moth. n.d. 'The Art and Craft of Storytelling'. http://themoth.org/. Accessed 4 October 2018.

Theatre of Debate. n.d. *Y Touring 21st Anniversary Interviews: Dr Martyn Thomas CBE FREng, Martyn Thomas Associates*. www.youtube.com/watch?v=SBNrTlCMgMk. Accessed 23 August 2018.

Thomas, Sophie, and Tim Williams. 2014. 'Khat (Catha Edulis): A Systematic Review of Evidence and Literature Pertaining to Its Harms to UK Users and Society'. *Drug Science, Policy and Law* 1 (April): 2050324513498332. https://doi.org/10.1177/2050324513498332.

Thompson, Charis. 2005. *Making Parents: The Ontological Choreography of Reproductive Technologies*. Cambridge, MA: MIT Press.

———. 2006. 'Race Science'. *Theory, Culture & Society* 23 (2–3): 547–9. https://doi.org/10.1177/026327640602300210 0.

Thompson, James. 2014. *Humanitarian Performance: From Disaster Tragedies to Spectacles of War*. Calcutta: Seagull Books.

Tierney, John. 2018. 'Dancing with the Scientists'. *The New York Times*, 18 September. https://tierneylab.blogs.nytimes.com/2008/11/20/dancing-with-the-scientists/.

Tönnies, Merle. 2017. 'The Immobility of Power in British Political Theatre after 2000: Absurdist Dystopias'. *Journal of Contemporary Drama in English; Berlin* 5 (1): 156–72. http://dx.doi.org/10.1515/jcde-2017-0012.

Toon, Elizabeth. 2014. 'The Machinery of Authoritarian Care: Dramatising Breast Cancer Treatment in 1970s Britain'. *Social History of Medicine* 27 (3): 557–76. https://doi.org/10.1093/shm/hku003.

Trafford, James, Robin Mackay, and Luke Pendrell. 2014. *Speculative Aesthetics*. Falmouth: Urbanomic.

Turing, A. M. 1950. 'Computing Machinery and Intelligence'. *Mind* LIX (236): 433–60. https://doi.org/10.1093/mind/LIX.236.433.

Turkle, Sherry. 2012. *Alone Together: Why We Expect More from Technology and Less from Each Other*. New York: Basic Books.

Turner, Victor. 1970. *The Forest of Symbols: Aspects of Ndembu Ritual*. Ithaca: Cornell University Press.

University of York Science Education Group. n.d. 'Twenty First Century Science Research-Informed Development'. University of York Department of Education. www.york.ac.uk/education/research/uyseg/projects/twentyfirstcenturyscience/twentyfirstcenturysciencresearch-informeddevelopment/. Accessed 20 June 2019.

Visvanathan, Shiv. 2003. 'Cultural Encounters and the Orient: A Study in the Politics of Knowledge'. *Diogenes* 50 (4): 69–81. https://doi.org/10.1177/03921921030504009.

———. 2005. 'Knowledge, Justice and Democracy'. In *Science and Citizens: Globalization and the Challenge of Engagement*, edited by Melissa Leach, Ian Scoones, and Brian Wynne, 83–94. London: Zed.

Wahlberg, Lars U. 2014. 'Chapter 62: Brain Implants'. In *Principles of Tissue Engineering*, edited by Robert Lanza, Robert Langer, and Joseph Vacanti, 1329–42. Boston: Academic Press. https://doi.org/10.1016/B978-0-12-398358-9.00062-8.

Wajcman, Judy. 2014. *Pressed for Time: The Acceleration of Life in Digital Capitalism*. Chicago: University of Chicago Press. http://ebookcentral.proquest.com/lib/manchester/detail.action?docID=3038639. Accessed 28 October 2019.

Walsh, Aylwyn, and Anna Ledgard. 2013. 'Re-Viewing an Arts-in-Health Process: For the Best'. *Research in Drama Education: The Journal of Applied Theatre and Performance* 18 (3): 216–29. https://doi.org/10.1080/13569783.2013.810923.

Warner, Michael. 2005. *Publics and Counterpublics*. New York: Zone Books.

Warsama, Yusra and Cora Bissett. 2015. *Rites*. London: Oberon Books.

Wartenberg, Thomas E. 1982. '"Species-Being" and "Human Nature" in Marx'. *Human Studies* 5 (2): 77–95. www.jstor.org/stable/20008832.

'Weekly Word Watch: Avozilla, Hooyah!, And #Merky'. 2018. OxfordWords Blog. 13 July. https://blog.oxforddictionaries.com/2018/07/13/weekly-word-watch-avozilla-hooyah-merky/. Accessed 17 May 2019.

Weibel, Peter, Ljiljana Fruk, and Zentrum für Kunst und Medientechnologie Karlsruhe. 2013. *Molecular Aesthetics*. Karlsruhe: ZKM/Center for Art and Media.

Weizenbaum, Joseph. 1976. *Computer Power and Human Reason: From Judgement to Calculation*. San Francisco: W.H. Freeman & Co Ltd.

Westfall, Richard. 1995. 'George Graham'. The Galileo Project. http://galileo.rice.edu/Catalog/NewFiles/graham.html. Accessed 28 June 2019.

White House Office of the Press Secretary. 2000. 'Remarks Made by the President, Prime Minister Tony Blair of England (via Satellite), Dr. Francis Collins, Director of the National Human Genome Research Institute, and Dr. Craig Venter, President and Chief Scientific Officer, Celera Genomics Corporation, on the Completion of the First Survey of the Entire Human Genome Project'. 26 June 2000. https://clintonwhitehouse3.archives.gov/WH/New/html/genome-20000626.html. Accessed 28 June 2019.

Williams, Simon. 2019. 'UK's First Earth System Model Launched'. Centre for Ecology & Hydrology. 11 March. www.ceh.ac.uk/news-and-media/news/uk-first-earth-system-model-launched. Accessed 26 November 2019.

Wolf, Stacy. 2007. '*Wicked* Divas, Musical Theater, and Internet Girl Fans'. *Camera Obscura: Feminism, Culture, and Media Studies* 22 (2 (65)): 39–71. https://doi.org/10.1215/02705346-2007-003.

———. 2008. '"Defying Gravity": Queer Conventions in the Musical *Wicked*'. *Theatre Journal* 60 (1): 1–21. https://doi.org/10.1353/tj.2008.0075.

Wolff, Tamsen. 2014. *Mendel's Theatre: Heredity, Eugenics, and Early Twentieth-Century American Drama*. New York: Palgrave Macmillan.

Wooster, Roger. 2016. *Theatre in Education in Britain: Origins, Development and Influence*. London: Bloomsbury Methuen Drama.

Young, Iris Marion. 2002. *Inclusion and Democracy*. Oxford: Oxford University Press.

Yusuf, Asha Lul Mohamud. 2012. 'Recollection'. Translated by Claire Pollard and Said Jama Hussein. Poetry Translation Centre. 2012. www.poetrytranslation.org/podcasts/recollection-by-asha-lul-mohamud-yusuf. Accessed 26 November 2019.

Zamyatin, Yevgeny. 1993. *We*. Translated by Clarence Brown. New York: Penguin Classics.

Zehelein, Eva-Sabine. 2009. *Science: Dramatic: Science Plays in America and Great Britain, 1990–2007*. Heidelberg: Universitätsverlag Winter.

Zemin, Jiang. 1998. 'Speech at the Sixth APEC Informal Leadership Meeting'. 18 November. www.fmprc.gov.cn/mfa_eng/wjdt_665385/zyjh_665391/t24917.shtml. Accessed 28 June 2019.

Zuger, Abigail. 2016. 'The Effect: A New Play Takes on Psychopharmacology, in All Its Human and Scientific Complexity'. *The Journal of the American Medical Association* 316 (14): 1434–5. https://doi.org/10.1001/jama.2016.14526.

Zupančič, Alenka. 2008. *The Odd One in: On Comedy*. Cambridge, MA: MIT Press.

———. 2016. '"You'd Have to Be Stupid Not to See That"'. *Parallax* 22 (4): 413–25. https://doi.org/10.1080/13534645.2016.1229164. Accessed 28 June 2019.

INDEX

Note: titles of theatre or dance performances and novels can be found under authors' names; films, other media and official reports are listed by title.

EU authorised representative for GPSR:
Easy Access System Europe, Mustamäe tee 50,
10621 Tallinn, Estonia
gpsr.requests@easproject.com

www.ingramcontent.com/pod-product-compliance
Ingram Content Group UK Ltd.
Pitfield, Milton Keynes, MK11 3LW, UK
UKHW021833100926
8556IPUK00004B/321